Algebra: A Very Short Introduction

VERY SHORT INTRODUCTIONS are for anyone wanting a stimulating and accessible way into a new subject. They are written by experts, and have been translated into more than 45 different languages.

The series began in 1995, and now covers a wide variety of topics in every discipline. The VSI library now contains over 500 volumes—a Very Short Introduction to everything from Psychology and Philosophy of Science to American History and Relativity—and continues to grow in every subject area.

Titles in the series include the following:

ACCOUNTING Christopher Nobes
ADOLESCENCE Peter K. Smith
ADVERTISING Winston Fletcher
AFRICAN AMERICAN RELIGION Eddie S. Glaude Jr
AFRICAN HISTORY John Parker and Richard Rathbone
AFRICAN RELIGIONS Jacob K. Olupona
AGEING Nancy A. Pachana
AGNOSTICISM Robin Le Poidevin
AGRICULTURE Paul Brassley and Richard Soffe
ALEXANDER THE GREAT Hugh Bowden
ALGEBRA Peter M. Higgins
AMERICAN HISTORY Paul S. Boyer
AMERICAN IMMIGRATION David A. Gerber
AMERICAN LEGAL HISTORY G. Edward White
AMERICAN POLITICAL HISTORY Donald Critchlow
AMERICAN POLITICAL PARTIES AND ELECTIONS L. Sandy Maisel
AMERICAN POLITICS Richard M. Valelly
THE AMERICAN PRESIDENCY Charles O. Jones
THE AMERICAN REVOLUTION Robert J. Allison
AMERICAN SLAVERY Heather Andrea Williams
THE AMERICAN WEST Stephen Aron
AMERICAN WOMEN'S HISTORY Susan Ware
ANAESTHESIA Aidan O'Donnell
ANARCHISM Colin Ward
ANCIENT ASSYRIA Karen Radner
ANCIENT EGYPT Ian Shaw
ANCIENT EGYPTIAN ART AND ARCHITECTURE Christina Riggs
ANCIENT GREECE Paul Cartledge
THE ANCIENT NEAR EAST Amanda H. Podany
ANCIENT PHILOSOPHY Julia Annas
ANCIENT WARFARE Harry Sidebottom
ANGELS David Albert Jones
ANGLICANISM Mark Chapman
THE ANGLO-SAXON AGE John Blair
ANIMAL BEHAVIOUR Tristram D. Wyatt
THE ANIMAL KINGDOM Peter Holland
ANIMAL RIGHTS David DeGrazia
THE ANTARCTIC Klaus Dodds
ANTISEMITISM Steven Beller
ANXIETY Daniel Freeman and Jason Freeman
THE APOCRYPHAL GOSPELS Paul Foster
ARCHAEOLOGY Paul Bahn
ARCHITECTURE Andrew Ballantyne
ARISTOCRACY William Doyle
ARISTOTLE Jonathan Barnes
ART HISTORY Dana Arnold
ART THEORY Cynthia Freeland

ASIAN AMERICAN HISTORY Madeline Y. Hsu
ASTROBIOLOGY David C. Catling
ASTROPHYSICS James Binney
ATHEISM Julian Baggini
THE ATMOSPHERE Paul I. Palmer
AUGUSTINE Henry Chadwick
AUSTRALIA Kenneth Morgan
AUTISM Uta Frith
THE AVANT GARDE David Cottington
THE AZTECS Davíd Carrasco
BABYLONIA Trevor Bryce
BACTERIA Sebastian G. B. Amyes
BANKING John Goddard and John O. S. Wilson
BARTHES Jonathan Culler
THE BEATS David Sterritt
BEAUTY Roger Scruton
BEHAVIOURAL ECONOMICS Michelle Baddeley
BESTSELLERS John Sutherland
THE BIBLE John Riches
BIBLICAL ARCHAEOLOGY Eric H. Cline
BIOGRAPHY Hermione Lee
BLACK HOLES Katherine Blundell
BLOOD Chris Cooper
THE BLUES Elijah Wald
THE BODY Chris Shilling
THE BOOK OF MORMON Terryl Givens
BORDERS Alexander C. Diener and Joshua Hagen
THE BRAIN Michael O'Shea
THE BRICS Andrew F. Cooper
THE BRITISH CONSTITUTION Martin Loughlin
THE BRITISH EMPIRE Ashley Jackson
BRITISH POLITICS Anthony Wright
BUDDHA Michael Carrithers
BUDDHISM Damien Keown
BUDDHIST ETHICS Damien Keown
BYZANTIUM Peter Sarris
CALVINISM Jon Balserak
CANCER Nicholas James
CAPITALISM James Fulcher
CATHOLICISM Gerald O'Collins
CAUSATION Stephen Mumford and Rani Lill Anjum
THE CELL Terence Allen and Graham Cowling
THE CELTS Barry Cunliffe
CHAOS Leonard Smith
CHEMISTRY Peter Atkins
CHILD PSYCHOLOGY Usha Goswami
CHILDREN'S LITERATURE Kimberley Reynolds
CHINESE LITERATURE Sabina Knight
CHOICE THEORY Michael Allingham
CHRISTIAN ART Beth Williamson
CHRISTIAN ETHICS D. Stephen Long
CHRISTIANITY Linda Woodhead
CIRCADIAN RHYTHMS Russell Foster and Leon Kreitzman
CITIZENSHIP Richard Bellamy
CIVIL ENGINEERING David Muir Wood
CLASSICAL LITERATURE William Allan
CLASSICAL MYTHOLOGY Helen Morales
CLASSICS Mary Beard and John Henderson
CLAUSEWITZ Michael Howard
CLIMATE Mark Maslin
CLIMATE CHANGE Mark Maslin
CLINICAL PSYCHOLOGY Susan Llewelyn and Katie Aafjes-van Doorn
COGNITIVE NEUROSCIENCE Richard Passingham
THE COLD WAR Robert McMahon
COLONIAL AMERICA Alan Taylor
COLONIAL LATIN AMERICAN LITERATURE Rolena Adorno
COMBINATORICS Robin Wilson
COMEDY Matthew Bevis
COMMUNISM Leslie Holmes
COMPLEXITY John H. Holland
THE COMPUTER Darrel Ince
COMPUTER SCIENCE Subrata Dasgupta
CONFUCIANISM Daniel K. Gardner
THE CONQUISTADORS Matthew Restall and Felipe Fernández-Armesto
CONSCIENCE Paul Strohm
CONSCIOUSNESS Susan Blackmore
CONTEMPORARY ART Julian Stallabrass
CONTEMPORARY FICTION Robert Eaglestone
CONTINENTAL PHILOSOPHY Simon Critchley
COPERNICUS Owen Gingerich
CORAL REEFS Charles Sheppard

CORPORATE SOCIAL RESPONSIBILITY Jeremy Moon
CORRUPTION Leslie Holmes
COSMOLOGY Peter Coles
CRIME FICTION Richard Bradford
CRIMINAL JUSTICE Julian V. Roberts
CRITICAL THEORY Stephen Eric Bronner
THE CRUSADES Christopher Tyerman
CRYPTOGRAPHY Fred Piper and Sean Murphy
CRYSTALLOGRAPHY A. M. Glazer
THE CULTURAL REVOLUTION Richard Curt Kraus
DADA AND SURREALISM David Hopkins
DANTE Peter Hainsworth and David Robey
DARWIN Jonathan Howard
THE DEAD SEA SCROLLS Timothy Lim
DECOLONIZATION Dane Kennedy
DEMOCRACY Bernard Crick
DEPRESSION Jan Scott and Mary Jane Tacchi
DERRIDA Simon Glendinning
DESCARTES Tom Sorell
DESERTS Nick Middleton
DESIGN John Heskett
DEVELOPMENTAL BIOLOGY Lewis Wolpert
THE DEVIL Darren Oldridge
DIASPORA Kevin Kenny
DICTIONARIES Lynda Mugglestone
DINOSAURS David Norman
DIPLOMACY Joseph M. Siracusa
DOCUMENTARY FILM Patricia Aufderheide
DREAMING J. Allan Hobson
DRUGS Les Iversen
DRUIDS Barry Cunliffe
EARLY MUSIC Thomas Forrest Kelly
THE EARTH Martin Redfern
EARTH SYSTEM SCIENCE Tim Lenton
ECONOMICS Partha Dasgupta
EDUCATION Gary Thomas
EGYPTIAN MYTH Geraldine Pinch
EIGHTEENTH-CENTURY BRITAIN Paul Langford
THE ELEMENTS Philip Ball
EMOTION Dylan Evans
EMPIRE Stephen Howe
ENGELS Terrell Carver
ENGINEERING David Blockley
ENGLISH LITERATURE Jonathan Bate
THE ENLIGHTENMENT John Robertson
ENTREPRENEURSHIP Paul Westhead and Mike Wright
ENVIRONMENTAL ECONOMICS Stephen Smith
ENVIRONMENTAL POLITICS Andrew Dobson
EPICUREANISM Catherine Wilson
EPIDEMIOLOGY Rodolfo Saracci
ETHICS Simon Blackburn
ETHNOMUSICOLOGY Timothy Rice
THE ETRUSCANS Christopher Smith
EUGENICS Philippa Levine
THE EUROPEAN UNION John Pinder and Simon Usherwood
EVOLUTION Brian and Deborah Charlesworth
EXISTENTIALISM Thomas Flynn
EXPLORATION Stewart A. Weaver
THE EYE Michael Land
FAMILY LAW Jonathan Herring
FASCISM Kevin Passmore
FASHION Rebecca Arnold
FEMINISM Margaret Walters
FILM Michael Wood
FILM MUSIC Kathryn Kalinak
THE FIRST WORLD WAR Michael Howard
FOLK MUSIC Mark Slobin
FOOD John Krebs
FORENSIC PSYCHOLOGY David Canter
FORENSIC SCIENCE Jim Fraser
FORESTS Jaboury Ghazoul
FOSSILS Keith Thomson
FOUCAULT Gary Gutting
THE FOUNDING FATHERS R. B. Bernstein
FRACTALS Kenneth Falconer
FREE SPEECH Nigel Warburton
FREE WILL Thomas Pink
FRENCH LITERATURE John D. Lyons
THE FRENCH REVOLUTION William Doyle
FREUD Anthony Storr
FUNDAMENTALISM Malise Ruthven

FUNGI Nicholas P. Money
THE FUTURE Jennifer M. Gidley
GALAXIES John Gribbin
GALILEO Stillman Drake
GAME THEORY Ken Binmore
GANDHI Bhikhu Parekh
GENES Jonathan Slack
GENIUS Andrew Robinson
GEOGRAPHY John Matthews and David Herbert
GEOPOLITICS Klaus Dodds
GERMAN LITERATURE Nicholas Boyle
GERMAN PHILOSOPHY Andrew Bowie
GLOBAL CATASTROPHES Bill McGuire
GLOBAL ECONOMIC HISTORY Robert C. Allen
GLOBALIZATION Manfred Steger
GOD John Bowker
GOETHE Ritchie Robertson
THE GOTHIC Nick Groom
GOVERNANCE Mark Bevir
GRAVITY Timothy Clifton
THE GREAT DEPRESSION AND THE NEW DEAL Eric Rauchway
HABERMAS James Gordon Finlayson
THE HABSBURG EMPIRE Martyn Rady
HAPPINESS Daniel M. Haybron
THE HARLEM RENAISSANCE Cheryl A. Wall
THE HEBREW BIBLE AS LITERATURE Tod Linafelt
HEGEL Peter Singer
HEIDEGGER Michael Inwood
HERMENEUTICS Jens Zimmermann
HERODOTUS Jennifer T. Roberts
HIEROGLYPHS Penelope Wilson
HINDUISM Kim Knott
HISTORY John H. Arnold
THE HISTORY OF ASTRONOMY Michael Hoskin
THE HISTORY OF CHEMISTRY William H. Brock
THE HISTORY OF LIFE Michael Benton
THE HISTORY OF MATHEMATICS Jacqueline Stedall
THE HISTORY OF MEDICINE William Bynum
THE HISTORY OF TIME Leofranc Holford-Strevens
HIV AND AIDS Alan Whiteside
HOBBES Richard Tuck
HOLLYWOOD Peter Decherney
HOME Michael Allen Fox
HORMONES Martin Luck
HUMAN ANATOMY Leslie Klenerman
HUMAN EVOLUTION Bernard Wood
HUMAN RIGHTS Andrew Clapham
HUMANISM Stephen Law
HUME A. J. Ayer
HUMOUR Noël Carroll
THE ICE AGE Jamie Woodward
IDEOLOGY Michael Freeden
INDIAN CINEMA Ashish Rajadhyaksha
INDIAN PHILOSOPHY Sue Hamilton
THE INDUSTRIAL REVOLUTION Robert C. Allen
INFECTIOUS DISEASE Marta L. Wayne and Benjamin M. Bolker
INFINITY Ian Stewart
INFORMATION Luciano Floridi
INNOVATION Mark Dodgson and David Gann
INTELLIGENCE Ian J. Deary
INTELLECTUAL PROPERTY Siva Vaidhyanathan
INTERNATIONAL LAW Vaughan Lowe
INTERNATIONAL MIGRATION Khalid Koser
INTERNATIONAL RELATIONS Paul Wilkinson
INTERNATIONAL SECURITY Christopher S. Browning
IRAN Ali M. Ansari
ISLAM Malise Ruthven
ISLAMIC HISTORY Adam Silverstein
ISOTOPES Rob Ellam
ITALIAN LITERATURE Peter Hainsworth and David Robey
JESUS Richard Bauckham
JOURNALISM Ian Hargreaves
JUDAISM Norman Solomon
JUNG Anthony Stevens
KABBALAH Joseph Dan
KAFKA Ritchie Robertson
KANT Roger Scruton
KEYNES Robert Skidelsky
KIERKEGAARD Patrick Gardiner
KNOWLEDGE Jennifer Nagel

THE KORAN Michael Cook
LANDSCAPE ARCHITECTURE
    Ian H. Thompson
LANDSCAPES AND
    GEOMORPHOLOGY
    Andrew Goudie and Heather Viles
LANGUAGES Stephen R. Anderson
LATE ANTIQUITY Gillian Clark
LAW Raymond Wacks
THE LAWS OF THERMODYNAMICS
    Peter Atkins
LEADERSHIP Keith Grint
LEARNING Mark Haselgrove
LEIBNIZ Maria Rosa Antognazza
LIBERALISM Michael Freeden
LIGHT Ian Walmsley
LINCOLN Allen C. Guelzo
LINGUISTICS Peter Matthews
LITERARY THEORY Jonathan Culler
LOCKE John Dunn
LOGIC Graham Priest
LOVE Ronald de Sousa
MACHIAVELLI Quentin Skinner
MADNESS Andrew Scull
MAGIC Owen Davies
MAGNA CARTA Nicholas Vincent
MAGNETISM Stephen Blundell
MALTHUS Donald Winch
MANAGEMENT John Hendry
MAO Delia Davin
MARINE BIOLOGY Philip V. Mladenov
THE MARQUIS DE SADE John Phillips
MARTIN LUTHER Scott H. Hendrix
MARTYRDOM Jolyon Mitchell
MARX Peter Singer
MATERIALS Christopher Hall
MATHEMATICS Timothy Gowers
THE MEANING OF LIFE
    Terry Eagleton
MEASUREMENT David Hand
MEDICAL ETHICS Tony Hope
MEDICAL LAW Charles Foster
MEDIEVAL BRITAIN John Gillingham
    and Ralph A. Griffiths
MEDIEVAL LITERATURE
    Elaine Treharne
MEDIEVAL PHILOSOPHY
    John Marenbon
MEMORY Jonathan K. Foster
METAPHYSICS Stephen Mumford

THE MEXICAN REVOLUTION
    Alan Knight
MICHAEL FARADAY Frank A. J. L. James
MICROBIOLOGY Nicholas P. Money
MICROECONOMICS Avinash Dixit
MICROSCOPY Terence Allen
THE MIDDLE AGES Miri Rubin
MILITARY JUSTICE Eugene R. Fidell
MINERALS David Vaughan
MODERN ART David Cottington
MODERN CHINA Rana Mitter
MODERN DRAMA
    Kirsten E. Shepherd-Barr
MODERN FRANCE Vanessa R. Schwartz
MODERN IRELAND Senia Pašeta
MODERN ITALY Anna Cento Bull
MODERN JAPAN
    Christopher Goto-Jones
MODERN LATIN AMERICAN
    LITERATURE
    Roberto González Echevarría
MODERN WAR Richard English
MODERNISM Christopher Butler
MOLECULAR BIOLOGY Aysha Divan
    and Janice A. Royds
MOLECULES Philip Ball
THE MONGOLS Morris Rossabi
MOONS David A. Rothery
MORMONISM
    Richard Lyman Bushman
MOUNTAINS Martin F. Price
MUHAMMAD Jonathan A. C. Brown
MULTICULTURALISM Ali Rattansi
MUSIC Nicholas Cook
MYTH Robert A. Segal
THE NAPOLEONIC WARS
    Mike Rapport
NATIONALISM Steven Grosby
NAVIGATION Jim Bennett
NELSON MANDELA Elleke Boehmer
NEOLIBERALISM Manfred Steger and
    Ravi Roy
NETWORKS Guido Caldarelli and
    Michele Catanzaro
THE NEW TESTAMENT
    Luke Timothy Johnson
THE NEW TESTAMENT AS
    LITERATURE Kyle Keefer
NEWTON Robert Iliffe
NIETZSCHE Michael Tanner

NINETEENTH-CENTURY BRITAIN
    Christopher Harvie and
    H. C. G. Matthew
THE NORMAN CONQUEST
    George Garnett
NORTH AMERICAN INDIANS
    Theda Perdue and Michael D. Green
NORTHERN IRELAND
    Marc Mulholland
NOTHING Frank Close
NUCLEAR PHYSICS Frank Close
NUCLEAR POWER Maxwell Irvine
NUCLEAR WEAPONS
    Joseph M. Siracusa
NUMBERS Peter M. Higgins
NUTRITION David A. Bender
OBJECTIVITY Stephen Gaukroger
THE OLD TESTAMENT
    Michael D. Coogan
THE ORCHESTRA D. Kern Holoman
ORGANIC CHEMISTRY
    Graham Patrick
ORGANIZATIONS Mary Jo Hatch
PAGANISM Owen Davies
THE PALESTINIAN-ISRAELI
    CONFLICT Martin Bunton
PANDEMICS Christian W. McMillen
PARTICLE PHYSICS Frank Close
PAUL E. P. Sanders
PEACE Oliver P. Richmond
PENTECOSTALISM William K. Kay
THE PERIODIC TABLE Eric R. Scerri
PHILOSOPHY Edward Craig
PHILOSOPHY IN THE ISLAMIC
    WORLD Peter Adamson
PHILOSOPHY OF LAW
    Raymond Wacks
PHILOSOPHY OF SCIENCE
    Samir Okasha
PHOTOGRAPHY Steve Edwards
PHYSICAL CHEMISTRY Peter Atkins
PILGRIMAGE Ian Reader
PLAGUE Paul Slack
PLANETS David A. Rothery
PLANTS Timothy Walker
PLATE TECTONICS Peter Molnar
PLATO Julia Annas
POLITICAL PHILOSOPHY
    David Miller
POLITICS Kenneth Minogue
POPULISM Cas Mudde and
    Cristóbal Rovira Kaltwasser
POSTCOLONIALISM Robert Young
POSTMODERNISM Christopher Butler
POSTSTRUCTURALISM
    Catherine Belsey
PREHISTORY Chris Gosden
PRESOCRATIC PHILOSOPHY
    Catherine Osborne
PRIVACY Raymond Wacks
PROBABILITY John Haigh
PROGRESSIVISM Walter Nugent
PROTESTANTISM Mark A. Noll
PSYCHIATRY Tom Burns
PSYCHOANALYSIS Daniel Pick
PSYCHOLOGY Gillian Butler and
    Freda McManus
PSYCHOTHERAPY Tom Burns and
    Eva Burns-Lundgren
PUBLIC ADMINISTRATION
    Stella Z. Theodoulou and Ravi K. Roy
PUBLIC HEALTH Virginia Berridge
PURITANISM Francis J. Bremer
THE QUAKERS Pink Dandelion
QUANTUM THEORY
    John Polkinghorne
RACISM Ali Rattansi
RADIOACTIVITY Claudio Tuniz
RASTAFARI Ennis B. Edmonds
THE REAGAN REVOLUTION Gil Troy
REALITY Jan Westerhoff
THE REFORMATION Peter Marshall
RELATIVITY Russell Stannard
RELIGION IN AMERICA Timothy Beal
THE RENAISSANCE Jerry Brotton
RENAISSANCE ART
    Geraldine A. Johnson
REVOLUTIONS Jack A. Goldstone
RHETORIC Richard Toye
RISK Baruch Fischhoff and John Kadvany
RITUAL Barry Stephenson
RIVERS Nick Middleton
ROBOTICS Alan Winfield
ROCKS Jan Zalasiewicz
ROMAN BRITAIN Peter Salway
THE ROMAN EMPIRE
    Christopher Kelly
THE ROMAN REPUBLIC
    David M. Gwynn
ROMANTICISM Michael Ferber

- ROUSSEAU Robert Wokler
- RUSSELL A. C. Grayling
- RUSSIAN HISTORY Geoffrey Hosking
- RUSSIAN LITERATURE Catriona Kelly
- THE RUSSIAN REVOLUTION S. A. Smith
- SAVANNAS Peter A. Furley
- SCHIZOPHRENIA Chris Frith and Eve Johnstone
- SCHOPENHAUER Christopher Janaway
- SCIENCE AND RELIGION Thomas Dixon
- SCIENCE FICTION David Seed
- THE SCIENTIFIC REVOLUTION Lawrence M. Principe
- SCOTLAND Rab Houston
- SEXUALITY Véronique Mottier
- SHAKESPEARE'S COMEDIES Bart van Es
- SIKHISM Eleanor Nesbitt
- THE SILK ROAD James A. Millward
- SLANG Jonathon Green
- SLEEP Steven W. Lockley and Russell G. Foster
- SOCIAL AND CULTURAL ANTHROPOLOGY John Monaghan and Peter Just
- SOCIAL PSYCHOLOGY Richard J. Crisp
- SOCIAL WORK Sally Holland and Jonathan Scourfield
- SOCIALISM Michael Newman
- SOCIOLINGUISTICS John Edwards
- SOCIOLOGY Steve Bruce
- SOCRATES C. C. W. Taylor
- SOUND Mike Goldsmith
- THE SOVIET UNION Stephen Lovell
- THE SPANISH CIVIL WAR Helen Graham
- SPANISH LITERATURE Jo Labanyi
- SPINOZA Roger Scruton
- SPIRITUALITY Philip Sheldrake
- SPORT Mike Cronin
- STARS Andrew King
- STATISTICS David J. Hand
- STEM CELLS Jonathan Slack
- STRUCTURAL ENGINEERING David Blockley
- STUART BRITAIN John Morrill
- SUPERCONDUCTIVITY Stephen Blundell
- SYMMETRY Ian Stewart
- TAXATION Stephen Smith
- TEETH Peter S. Ungar
- TELESCOPES Geoff Cottrell
- TERRORISM Charles Townshend
- THEATRE Marvin Carlson
- THEOLOGY David F. Ford
- THOMAS AQUINAS Fergus Kerr
- THOUGHT Tim Bayne
- TIBETAN BUDDHISM Matthew T. Kapstein
- TOCQUEVILLE Harvey C. Mansfield
- TRAGEDY Adrian Poole
- TRANSLATION Matthew Reynolds
- THE TROJAN WAR Eric H. Cline
- TRUST Katherine Hawley
- THE TUDORS John Guy
- TWENTIETH-CENTURY BRITAIN Kenneth O. Morgan
- THE UNITED NATIONS Jussi M. Hanhimäki
- THE U.S. CONGRESS Donald A. Ritchie
- THE U.S. SUPREME COURT Linda Greenhouse
- UTOPIANISM Lyman Tower Sargent
- THE VIKINGS Julian Richards
- VIRUSES Dorothy H. Crawford
- VOLTAIRE Nicholas Cronk
- WAR AND TECHNOLOGY Alex Roland
- WATER John Finney
- WEATHER Storm Dunlop
- THE WELFARE STATE David Garland
- WILLIAM SHAKESPEARE Stanley Wells
- WITCHCRAFT Malcolm Gaskill
- WITTGENSTEIN A. C. Grayling
- WORK Stephen Fineman
- WORLD MUSIC Philip Bohlman
- THE WORLD TRADE ORGANIZATION Amrita Narlikar
- WORLD WAR II Gerhard L. Weinberg
- WRITING AND SCRIPT Andrew Robinson
- ZIONISM Michael Stanislawski

Peter M. Higgins

ALGEBRA

A Very Short Introduction

OXFORD
UNIVERSITY PRESS

Great Clarendon Street, Oxford, OX2 6DP,
United Kingdom

Oxford University Press is a department of the University of Oxford.
It furthers the University's objective of excellence in research, scholarship,
and education by publishing worldwide. Oxford is a registered trade mark of
Oxford University Press in the UK and in certain other countries

© Peter M. Higgins 2015

The moral rights of the author have been asserted

First edition published in 2015

Impression: 3

All rights reserved. No part of this publication may be reproduced, stored in
a retrieval system, or transmitted, in any form or by any means, without the
prior permission in writing of Oxford University Press, or as expressly permitted
by law, by licence or under terms agreed with the appropriate reprographics
rights organization. Enquiries concerning reproduction outside the scope of the
above should be sent to the Rights Department, Oxford University Press, at the
address above

You must not circulate this work in any other form
and you must impose this same condition on any acquirer

Published in the United States of America by Oxford University Press
198 Madison Avenue, New York, NY 10016, United States of America

British Library Cataloguing in Publication Data

Data available

Library of Congress Control Number: 2015938927

ISBN 978-0-19-873282-2

Printed in Great Britain by
Ashford Colour Press Ltd, Gosport, Hampshire

Contents

Preface xiii

List of illustrations xv

1 Numbers and algebra 1

2 The laws of algebra 11

3 Linear equations and inequalities 25

4 Quadratic equations 40

5 The algebra of polynomials and cubic equations 53

6 Algebra and the arithmetic of remainders 75

7 Introduction to matrices 87

8 Matrices and groups 101

9 Determinants and matrices 111

10 Vector spaces 126

Further reading 139

Index 141

Preface

Algebra is the lingua franca of the mathematical sciences and the purpose of this little book is to explain what it is about. The laws governing algebra emerge from the behaviour of numbers, and one of our themes is how many of the rules and practices of arithmetic and algebra are consequences of a small collection of fundamental laws that represent familiar properties of ordinary integers.

The first half of the book establishes much of the algebra that has been a staple of secondary school mathematics for generations, which is based on finding unknowns in linear and quadratic equations, and this the reader meets in the first four chapters. Modern algebra was born out of the struggle to solve equations of degree higher than 2, and the first part of the book culminates in Chapter 5 with finding solutions of general cubic equations, the roots of which are not necessarily just simple fractions but may involve so-called irrational and complex numbers.

The second half of the book introduces modern aspects of the subject and we look at algebra that is not based on the general behaviour of numbers but involves other kinds of mathematical objects. The topic of Chapter 6 is the arithmetic of remainders, which furnishes examples of a fundamental algebra type with two operations, namely that of a ring. Matrices are the central feature of Chapters 7, 8, and 9. The origin of matrices may be traced back thousands of years to ancient China but the topic only gained traction in the middle of the 19th century, from which point it has grown to become the primary vehicle for calculation throughout mathematics, physics, and the social sciences. The historical significance of matrix theory in pure mathematics, however, is that it provided an important example of another type of algebra apart from number fields. The final chapter introduces vector spaces and finite fields.

Many aspects of algebra are touched upon in the course of the book and every piece matters. My hope is that readers will see the parts of the jigsaw come together as they move through the book and thereby gain an appreciation of algebra as a whole. Modern abstract algebra is firmly based on what are known as groups, rings, fields, and vector spaces. The reader is made aware of these constructs through examples that emerge in the development of the text. Only after that are these ideas introduced in a more formal fashion. The intention is that the reader will be left with both an overview of elementary mathematics and a taste for and an insight into contemporary aspects of the vast world of algebra.

Peter M. Higgins,
Colchester, 2015

List of illustrations

1. Addition and subtraction on the number line **5**
2. Area of the inscribed rectangle = ? **17**
3. Graph of a typical linear function $y = ax + b$ **27**
4. Temperature in Celsius and Fahrenheit **29**
5. Signs of the expressions $x + 2$ and $3x + 5$ **34**
6. Comparing the graphs of $y = x^2$ and $y = x^2 + 6x + 13$ **44**
7. Graph of $y = x - x^2$ **45**
8. Golden Rectangle **49**
9. Three quadratic graphs exhibiting 0, 1, and 2 roots, respectively **50**
10. Man running for a bus **51**
11. The Argand plane with addition and subtraction of complex numbers **58**
12. Multiplication of complex numbers in polar form **63**
13. Network of cities and airline routes **93**
14. Geometric action of a matrix transformation **96**
15. Matrix acting on the unit square **99**

Chapter 1
Numbers and algebra

The backstory of algebra

In our schooldays, the arrival of x and y on the scene represented the point where mathematics went beyond mere arithmetic and, by acquiring a language all of its own, entered a higher realm. By passing through the portal of algebra, the subject develops surprising power, showing us things that could not be discovered in any other way. Modern science is based on mathematics, which comes about through algebraic manipulation of symbols representing quantities of interest. Algebra is the tool through which exact physical relationships are revealed, including that most famous of equations, $E = mc^2$, along with a host of others. Equations like this one, which arises in Einstein's theory of special relativity, are consequences of physical models based on experiment. Nonetheless, the relationship itself is arrived at through algebra, and it is the undoubted soundness of the underlying algebra that gives authority to the momentous conclusion that energy and mass are one and the same thing. Algebra underpins all modern systematic research. Although its contribution may be embedded in scientific software, without algebra, progress would be impossible.

The word 'algebra' is derived from the Arabic word *al-gebr*, meaning reunion of broken parts. During the 11th century, it was perhaps the Islamic world that represented the most mathematically sophisticated civilization. However, there was no algebraic manipulation of the kind seen in modern texts, and medieval mathematical writing throughout the world of Marco Polo was rhetorical, with everything being described in words. Algebra of a kind that we might recognize did not appear until the 17th century. The scarcity of paper may have held back the spontaneous development of mathematical symbology, but it should be appreciated also that ancient scholars faced obstacles that obscured the underlying mathematical landscape of arithmetic.

When we carry out algebraic manipulation we introduce arbitrary symbols, x and y being the most common, which stand for fixed but unspecified numbers, and these symbols are manipulated according to the laws of arithmetic. The argument underpinning all we do is that, no matter what the numbers x and y may be, the relationships that emerge from our manipulations are true as they are consequences of our initial assumptions and of rules that apply to all numbers, independently of their particular values. The use of algebraic symbols to stand for unknown quantities is a convenient abbreviation, and although that brevity certainly facilitates reasoning, the real power of algebra stems from the universality of interpretation that the symbols afford, which allows them be wielded in a powerful manner that cannot be matched by words alone.

In order to realize the potential of algebra, we need to be able to move our symbols about in an uninhibited fashion, making free use of the operations of arithmetic, particularly the fundamental pairs of operations: addition and subtraction, multiplication and division. For that we need a number system fit for purpose. If, for example, we reject negative quantities as meaningless or, more fundamentally still, we fail to treat zero as a number, we will be handicapped and deny ourselves the freedom that algebra offers to explore the world of unknown quantities. A cloud of confusion needed to be dissipated before the existence of the algebraic world that we take for granted could even be glimpsed, never mind properly understood and developed.

Great minds of the past would have been stunned at the ease with which a modern student can use algebra to completely solve problems that they found impossible and perhaps even had difficulty formulating with any clarity. For example, school algebra is enough to prove that the square root of a whole number, for example $\sqrt{36}$ or $\sqrt{42}$, either is another whole number or is not a fraction at all. The ancient Greek scholars put great effort into this question and used their geometric methods to show that some particular square roots up to $\sqrt{17}$ were not fractions. The general problem, however, defeated them, yet this and many others beyond the reach of the ancients can be completely understood by the reader of an OUP *Very Short Introduction*, as you shall see.

The number system

In order to harness the power of algebra we need a number system that meets its demands. Part of those demands is the freedom to perform the four basic arithmetic operations with symbols standing for arbitrary or unknown numbers. However, the collection of ordinary counting numbers has shortcomings in this regard. The numbers that arise through counting, 1, 2, 3, . . . are known as the *natural numbers* because they emerge more or less of their own accord once we begin to tally things up. This set of numbers is denoted by $\mathbb{N}$, and $\mathbb{N}$ is *closed* under the operations of addition and multiplication, meaning that if we begin with two natural numbers we may add or multiply them together and the answer is always another natural number. Subtraction, however, is a different story. Subtraction is the taking away of one number from another and is the reverse or, as mathematicians prefer to say, the *inverse* operation to addition. Applying subtraction in sums such as $3 - 5$ where the second number is larger that the first takes us out of $\mathbb{N}$ and into the realm of the *negative integers*, as they are called. When this kind of difficulty arises, we do not give up but rather adopt the attitude that our number system is currently inadequate and should be extended to allow continuation of our calculations.

The standard model of numbers that pervades all of advanced mathematics and engineering is the field of so-called complex numbers, denoted by $\mathbb{C}$. The journey from $\mathbb{N}$ all the way to $\mathbb{C}$ was long and was not truly completed until the 19th century. Prior to that there was much philosophical agonizing as to the reality, meaning, and validity of numbers other than the natural numbers. We shall, however, introduce the required number types without hesitation.

Having said this, we begin by adjoining to $\mathbb{N}$ the number zero, denoted by 0, which may be added to or subtracted from any number without changing its value. It must be conceded that 0 is not a member of the *positive integers*, as the natural numbers are sometimes called, but 0 is a number nonetheless and will need to find its place in our system of arithmetic. We next introduce a negative mirror image for each positive number; for instance, -6 is the negative partner for 6.

Although not necessary for the development of the subject, it is often easiest to picture and explain the behaviour of numbers by imagining them sitting along the *number line*. This is a horizontal line with the integers placed at equally spaced points along its length. We place 0 in the middle, the positive integers marching off to the right in their natural ascending order and the negatives occupying the mirror positions to the left of zero.

The collection of all *integers*, as this set is called, positive, negative, and zero, is represented by the symbol $\mathbb{Z}$, while $\mathbb{Q}$ stands for the collection of *rational numbers*, which comprise all fractions together with their negatives. The set $\mathbb{Z}$ lies within $\mathbb{Q}$, as an integer n is equal to the rational number $n/1$. (We say that $\mathbb{Z}$ is a *subset* of $\mathbb{Q}$, and, in like manner, $\mathbb{N}$ is a subset of $\mathbb{Z}$.) However, two rational numbers such as 3/9 and 7/21 are considered to be equal as they both cancel down to the same fraction, in this case 1/3. Any positive rational number has a unique representation as a fraction cancelled to *lowest terms*, a/b, where a and b have no common factor other than 1. The rational numbers too may be pictured as lying along the number line in their natural order, densely and uniformly spread throughout its length.

To add a positive number n to another number m (positive or not), we begin at m and move n places to the right on the number line, while to subtract n we move n places to the left. In the set $\mathbb{Z}$, each number n has an opposite, $-n$, and we now use this feature to define addition of negatives in terms of subtraction. We declare that subtraction of any number n is to mean the same thing as the adding of its opposite, $-n$, so that adding a negative number $-n$ moves us n places to the *left* on the number line. It follows that to *subtract a negative number* $-n$, we add its opposite, n. In other words, to subtract the negative number $-n$ we move n places to the *right* on the number line.

This way of looking at things leads to familiar sums such as

$$(-1) + 4 = 3, \quad 6 + (-11) = -5, \quad (-8) + 6 = -2, \quad 1 - (-9) = 1 + 9 = 10,$$

as pictured in Figure 1.

The brackets around -1 and other negatives here are not strictly necessary but are introduced to avoid either beginning a string

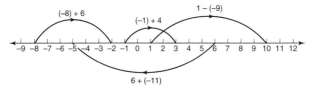

1. Addition and subtraction on the number line.

with an operation symbol or a clash of two operation symbols, for instance + and −. The need to do this comes about because we have loaded the minus symbol with two slightly different meanings: the minus sign is used both to indicate the taking of the opposite of an integer, which is an operation on a single number, and also to stand for subtraction, which is an operation on two numbers taken in a particular order.

Up to this point, we have not invoked anything that you might call a Law of Algebra to explain how our arithmetic works. The justification for our rules depends, rather, on extending the idea of subtraction to the entire collection of integers, which has been ordered in a natural linear fashion. In Chapter 2, we explore the laws that govern arithmetic operations and explain how these rules are extended so that they continue to be respected as we pass from one number system to a greater one that subsumes the former.

Number factorization

Although division of one integer by another generally leads to a fractional answer that lies outside the integers, division of one whole number by another may have an integer outcome, and the nature of how and when this happens is important and finds analogues in other algebraic systems we shall meet, such as polynomials. For that reason, we now record the main features of integer division. We will begin to use both *power notation* (for instance, writing $2 \times 2 \times 2$ as 2^3) and the 'less than' and 'less than or equal to' signs, $<$ and $\leq$, respectively (for example, $4 < 7$ and $-3 < 2$, as in each instance the first number lies to the left of the second on the number line). When it is understood that we are dealing with multiplication of numbers represented by letters such as a and b, we normally take the multiplication sign as given and so write ab or sometimes $a \cdot b$ instead of $a \times b$. We tend to avoid the cumbersome $\times$ sign and sometimes write arithmetic expressions like $2 \times (-3) \times 4$ as $(2)(-3)(4)$.

An integer a is a *factor* or *divisor* of another integer b if b can be written as $b = ac$, where c is itself an integer (equally, of course, c is then also a factor of b). A *prime* is a positive integer such as 71 that has just two positive factors, those necessarily being 1 and the number itself. An integer exceeding 1 that is not prime is called *composite*, as it is composed of smaller factors. For example $72 = 8 \times 9$. We say that 8 is a factor of 72 or that 8 *divides* 72 or that 72 is a *multiple* of 8: we sometimes denote this relationship by 8|72, which is simply shorthand for '8 is a factor of 72'. Successively factorizing the divisors of a given number as far as possible will eventually yield the *prime factorization* of the number. In our example, $72 = 8 \times 9 = 2^3 \times 3^2$. We could have found the prime factorization of 72 by another route by writing $72 = 6 \times 12 = (2 \times 3) \times (4 \times 3) = (2 \times 3) \times (2 \times 2 \times 3)$, but rearranging the prime factors from lowest to highest yields the same result as before, and we say that $2^3 \times 3^2$ is *the* prime factorization of 72. The *Fundamental Theorem of Arithmetic* says that the prime factorization of any natural number n (with the prime factors written in ascending order) is unique. This uniqueness can be deduced from an even more basic property of numbers, *Euclid's Lemma*, which says that if a prime number p divides a product ab, so that $p|ab$, then p is a factor of a or a factor of b (or perhaps a factor of both). An equivalent formulation of Euclid's Lemma is that *if* neither a nor b is a multiple of the prime p, *then* nor is their product ab. Although plausible, this property is not self-evident, and we do not prove Euclid's Lemma here. We shall, however, explain more about why it holds later in this section. (My VSI *Numbers* explains in detail all the properties of integers that are taken for granted here.)

The general pattern that arises when one natural number, a, is divided by another, b, is as follows. To divide b into a, we subtract as many b's as we can from a, q say, until the *remainder* $r < b$. In this way, we get $a = bq + r$. This expression is unique: there is only one value for q and one for r that make this equation true, remembering that we are insisting that $0 \leq r \leq b - 1$. There are special cases, for instance $q = 0$ exactly when $a < b$, in which case $r = a$. More interestingly, $r = 0$ exactly when $b|a$, in which case $a/b = q$.

As a representative example, if $a = 72$ and $b = 13$ then $72 = 13 \times 5 + 7$, so here we have $q = 5$ and $r = 7$. This process of producing the equation $a = bq + r$ for given a and b is known as the *Division Algorithm*.

One fundamental algebraic idea that we first meet in arithmetic is that of the *greatest common divisor* (gcd), also known as the *highest common factor*, of two positive integers a and b. As the name seeks to convey, the gcd of a and b is the largest number d that is a factor of both a and b; since a and b always have at least one common factor, that being the number 1, the gcd certainly exists. We call two numbers a and b *relatively prime* to each other if their gcd is 1. For example, $15 = 3 \times 5$ and $28 = 2^2 \times 7$ are relatively prime (although neither number is itself prime). The question remains, however, as to how we may compute the gcd of two given numbers.

The gcd, d, can be found through comparison of the prime factorizations of a and b, for the prime factors of d are just those common to a and b. There is, however, a better way of finding it, known as the *Euclidean algorithm*, which not only is quicker but also reveals other useful relationships. We shall explain the algorithm shortly, but first we draw attention to certain basic properties of common factors.

Suppose that c is any common factor of a and b, so that $a = ct$ and $b = cs$, say. Then c is also a factor of any number r of the form $r = ax + by$, where x and y are themselves integers (which may be negative or zero). To show this we locate and 'take out' the common factor of c in the expression $ax + by$, as follows:

$$r = ax + by = ctx + csy = c(tx + sy). \tag{1}$$

Since $tx + sy$ is another integer, we have that c is indeed a factor of r.

An immediate consequence of (1) is that it applies to our Division Algorithm equation written in the form $r = a - bq$, for it tells us that any common factor of a and b is also a factor of r. By the same token, it follows from $a = bq + r$ that any common factor of b and r is also a factor of a. Hence the set of all common factors of a and b is the same as the set of all common factors of b and r and, in particular, the gcd of a and b is likewise the gcd of b and r. This allows us to work with the pair b and r instead of b and a and, since $r < b$, this simplifies our problem of finding the gcd as we can now apply the Division Algorithm to the pair (b, r) and repeat the process until the gcd of a and b emerges. This process is known as the Euclidean Algorithm.

Let us act with the algorithm on the pair $a = 189$ and $b = 105$. We underline the two numbers in hand at each stage and divide the smaller into the larger, discarding the larger as we proceed from one line to the next. We halt the procedure when the remainder becomes 0, indicating that the remainder on the previous line is the required gcd:

$$\underline{189} = 1 \times \underline{105} + 84,$$
$$\underline{105} = 1 \times \underline{84} + 21,$$
$$\underline{84} = 4 \times \underline{21},$$

and so the gcd of 189 and 105 is 21 ($189 = 9 \times 21$ and $105 = 5 \times 21$).

These equations themselves have uses as they can be reversed to express the gcd, d, in terms of the original numbers, a and b. We begin with the second last equation and make d the subject, giving in this case $21 = 105 - 84$. Then we use each equation in turn to eliminate an intermediate remainder: in our example, the first equation gives $84 = 189 - 105$ and so overall we have

$$21 = 105 - (189 - 105) = 2 \times 105 - 1 \times 189.$$

There are interesting theoretical consequences as well, which we will call upon in Chapter 6. We proved earlier in this section that any common factor c of a and b is also a factor of any number of the form $ax + by$, and since the gcd d of a and b has this form, which may be found by reversing the steps of the Euclidean Algorithm, it follows that *any* common factor c of a and b divides their gcd d. Moreover, $a' = a/d$ and $b' = b/d$ have a gcd of 1, for suppose that t is a common factor of a' and b' so that $a' = ta''$ and $b' = tb''$, say. We shall verify that $t = 1$. (The use of dashes is a way of reminding ourselves that a' and a'' are factors of a: of course, any new symbol could be used.) The previous equations imply that $a = da' = dta''$ and $b = db' = dtb''$, whence dt is a common factor of a and b. Since, however, d is the gcd of a and b, it follows that $t = 1$ and a' and b' are indeed relatively prime. In our example, $a = 189$, $b = 105$, and $d = 21$; dividing through by the gcd gives $189/21 = 9$ and $105/21 = 5$, and 9 and 5 have no common factor (apart from 1).

The fact that the gcd of two numbers a and b may be written in the form $ax + by$ lies at the heart of a host of algebraic proofs about numbers, Euclid's Lemma being just one of many examples.

Chapter 2
The laws of algebra

Rules, tricks, and traps

As mathematical novices we stumble through the world of algebra, suffering bumps and bruises along the way, much as we did when learning to walk. Almost by experiment we discover a list, often a rather long and disorderly list, of things that we can and cannot do when moving symbols about, and thereby form a collection of algebraic rules to live by.

In centuries past, it was worse, for the subject had no systematic foundation and it was not clear to what extent it was to be taken seriously. The 11th-century Persian mathematician and poet Omar Khayyam insisted that algebra was not merely a box of tricks for finding unknowns but rather an alternative path to geometrical truths. Despite this insight, he lacked anything resembling modern algebraic notation. Even negative numbers were regarded as formalisms at best, which held no intrinsic meaning but only mattered as part of some larger calculation.

Our approach here focuses on just three laws and we motivate them through our experience of working with the integers. First, however, a word about bracketing.

Brackets are used to indicate in which order the separate operations in an expression are to be carried out. When writing a sum like 3 + 7 + 8, however, we do not feel the need for brackets, as the two alternative ways of bracketing the expression lead to one and the same answer:

$$(3 + 7) + 8 = 10 + 8 = 18; \quad 3 + (7 + 8) = 3 + 15 = 18.$$

We say that the operation of addition is *associative*, meaning that $(a + b) + c = a + (b + c)$ always holds for arbitrary numbers a, b, and c. In the same way, we have the *associative law of multiplication*: $a(bc) = (ab)c$.

That numbers can be relied on to obey this pair of associative laws is not obvious, although it may be familiar. We shall, however, take this pair of laws for granted, although at the same time it should be appreciated that associativity fails for subtraction and for division. For example,

$$(11 - 6) - 3 = 5 - 3 = 2; \quad \text{but } 11 - (6 - 3) = 11 - 3 = 8.$$

You always have to take care when there are brackets and minus signs about. If we want to remove the brackets in the second sum we need to change the sign on each term in the bracket, and not just the first one, to get the correct result:

$$11 - (6 - 3) = 11 - 6 + 3 = 5 + 3 = 8.$$

Similarly, division is not associative either:

$$(24 \div 6) \div 2 = 4 \div 2 = 2; \text{ yet } 24 \div (6 \div 2) = 24 \div 3 = 8.$$

So different bracketings yield different outcomes.

There is a convention that sometimes avoids the need for bracketing: for example, when we write $11 - 6 - 3$, it is understood that this means $(11 - 6) - 3$, so that the operations are carried out in the order we meet them when working from left to right.

As a general principle, in the absence of bracketing directing us otherwise, the convention is that when there is a mix of multiplications and additions, multiplication takes precedence over addition, so for instance $4 + 7 \times 3$ implicitly means $4 + (7 \times 3) = 4 + 21 = 25$. If we required the addition to be performed first, we would need to communicate this through bracketing and write $(4 + 7) \times 3 = 11 \times 3 = 33$. We need to appreciate that $a + b \times c$ is just an abbreviated way of writing $a + (b \times c)$ and so this convention does *not* represent a law of algebra, since it does not assert a general fact about the nature of numbers.

There is a real law, however, that ties addition and multiplication together, and that is the *distributive law*—it's the one that tells us how to multiply out the brackets:

$$a(b + c) = ab + ac.$$

For positive a we can see why this law holds, as it is saying

$$\underbrace{(b+c) + (b+c) + \ldots + (b+c)}_{\times a} = \underbrace{(b + b + \ldots + b)}_{\times a} + \underbrace{(c + c + \ldots + c)}_{\times a},$$

which we can see comes about because when we add up a lots of b and a lots of c, the outcome is independent of the order in which the numbers are added. In saying this we are assuming another law, the *commutative law* of addition,

$$a + b = b + a,$$

a law that holds equally well for multiplication: $ab = ba$. These, then, are the laws that we will take forward with us: associativity and commutativity of addition and of multiplication, and the distributive law of addition over multiplication.

As a taste of what these rules are good for, we shall prove that $a \times 0 = 0$. Since $0 + 0 = 0$, we have by the distributive law that

$$a \times 0 = a \times (0 + 0) = (a \times 0) + (a \times 0).$$

Hence the number $b = a \times 0$ has the property that $b = b + b$. Subtracting b from both sides of this little equation now gives that $0 = b$, which is to say that $a \times 0 = 0$.

We are getting a little ahead of ourselves—we shall examine this kind of thing again later—but this argument does give an example of the way in which some properties of numbers are consequences of the laws of algebra as we have outlined them and do not have to be assumed.

Next we compare division with subtraction. We identify subtraction of a with the addition of its opposite, $-a$, where $-a$ is the number which when added to a gives zero: $a + (-a) = 0$. In algebra we call $-a$ the *inverse* of a with respect to addition. The number 0 is the *additive identity*, the number which when added to any other does not change its value: $a + 0 = a$. We duplicate this pattern with the operation of multiplication in order to define division in a consistent fashion.

First we identify the *multiplicative identity*, the number that when *multiplied* by any other number a does not alter its value. This number is of course the number 1: $a \times 1 = a$. The inverse of a number a is then its reciprocal, $1/a$, as $1/a$ has the property that $a \times 1/a = 1$. Analogously with subtraction, we now *define* division by a number a as multiplication by its inverse, $1/a$, so, for example, $8 \div (2/3) = 8 \times (3/2) = 4 \times 3 = 12$ as, for $a = 2/3, 1/a = 3/2$ because $(2/3) \times (3/2) = 1$. Division is the inverse operation to multiplication. And now we can explain another algebraic fact as well,

$$\frac{a+b}{c} = \frac{a}{c} + \frac{b}{c}, \qquad (2)$$

by pausing to contemplate what (2) actually says. Since dividing by c means multiplying by $1/c$, (2) can be written as

$$\frac{1}{c}(a+b) = \frac{1}{c} \cdot a + \frac{1}{c} \cdot b,$$

whereupon we see that rule (2) is nothing new but is rather just a special case of the distributive law.

When in doubt about what to do with an algebraic expression, a rule of thumb is 'put everything over a common denominator'. This step is based on the distributive law and is so commonly called upon that it needs to be made explicit. What this entails is a general description of what is done when two fractions are added:

$$\frac{a}{b} + \frac{c}{d};$$

the *denominators* (bottom lines) here are different, and that is a real incompatibility. These fractions use two different units of measurement: the first measures in terms of $1/b$, the second in terms of $1/d$. To make progress we need to express both fractions with a common denominator. The product bd is always a common multiple of b and d, so we continue in that way, using $1/bd$ as a common measure for both fractions, making use of our rule (2) in the reverse direction to complete the sum. However, whenever we multiply one of the denominators by some number, the same multiplication must be applied to the *numerator* (top line) as well so as not to change the value of the fraction. Following these dictums, the algebra goes as follows:

$$\frac{a}{b} + \frac{c}{d} = \frac{ad}{bd} + \frac{bc}{bd} = \frac{ad + bc}{bd}.$$

The distributive law also allows us to expand brackets involving sums of any length. A fundamental algebraic fact involves expansion of the square of a sum of two terms: $(a + b)^2$. Assigning a temporary symbol c to the first factor, $a + b$, we may harness the distributive and commutative laws together to obtain

$$(a + b)^2 = c(a + b) = ca + cb = ac + bc = a(a + b) + b(a + b)$$
$$= a^2 + ab + ba + b^2,$$

whereupon gathering together the two like terms in the centre gives

$$(a+b)^2 = a^2 + 2ab + b^2. \tag{3}$$

We call a result like (3) an *algebraic identity*, meaning that (3) is always true no matter which numbers are substituted for a and b. This is in contrast with an *equation*, such as $x^2 - 9x + 8 = 0$, which only holds for some values of x: in this case the equation is only valid if $x = 1$ or $x = 8$.

The operation of 'collecting together like terms' is again just an instance of the distributive law. For example, let's tidy up the algebraic expression

$$4a^2 - 5ab - a^2 + ba = 4a^2 - a^2 + ab - 5ab,$$

where we have rearranged the order of terms and within terms using commutativity. We continue with the 'collection' by employing the distributive law:

$$= a^2(4-1) + ab(1-5) = 3a^2 - 4ab.$$

Normally, of course, we do not record this level of detail but it should be appreciated that all common algebraic manipulations are, at bottom, justified by invoking the associative, commutative, and distributive laws.

A sister identity for (3) is arrived at for $(a-b)^2$ through writing c for $a-b$:

$$(a-b)^2 = c(a-b) = ac - bc = a(a-b) - b(a-b)$$
$$= a^2 - ab - ba + b^2,$$

and hence we obtain

$$(a-b)^2 = a^2 - 2ab + b^2.$$

We must not forget the presence of the *cross-terms*, $\pm 2ab$ (plus or minus $2ab$), in these expansions. In contrast, when we expand

$(a - b)(a + b)$ the cross-terms have opposite signs, thereby cancelling each other out. This yields an important expression for the *difference of two squares*:

$$a^2 - b^2 = (a - b)(a + b).$$

It is the use of the distributive law in this direction, where we express a sum or difference of terms as a product, which is particularly important. All throughout mathematics, whether we are dealing with numbers or with algebraic expressions, finding factorizations can be particularly useful. There is no end to the number and variety of factorization identities that can be discovered. Each of them can be routinely verified using the laws of algebra, but finding them in the first place is not such a mechanical process. As another sample, we list the sum and difference of two cubes:

$$a^3 + b^3 = (a + b)(a^2 - ab + b^2); \quad a^3 - b^3 = (a - b)(a^2 + ab + b^2).$$

A nice problem that exploits these ideas is the following: a rectangle of perimeter 28 is inscribed in a circle of radius 6. What is its area?

First, label the unknown sides of the rectangle as a and b, (Figure 2). The perimeter information is captured by the equation $2a + 2b = 28$, which gives $a + b = 14$. The diameter of the circle is $2 \times 6 = 12$ and so, by Pythagoras's Theorem, we have $a^2 + b^2 = 12^2$.

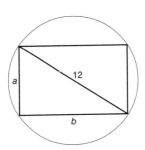

2. Area of the inscribed rectangle = ?

Finding a and b is now possible but is not required. We should keep our eyes on the prize—the question asks for the *area* of the rectangle, which is the product ab, and that may be found immediately using the identities just derived:

$(a + b)^2 = a^2 + b^2 + 2ab$ and so in particular $14^2 = 12^2 + 2ab$,

$$\therefore ab = \frac{1}{2}(14^2 - 12^2) = \frac{1}{2}(14 - 12)(14 + 12) = \frac{1}{2} \cdot 2 \cdot 26 = 26.$$

Powers, indices, and the Binomial Theorem

Since powers are beginning to emerge in our considerations, this is a suitable juncture to explain how they operate in general. The superscript in a power such as a^5 is called its *index*. The *first law of indices* is that $a^m \times a^n = a^{m+n}$. This is merely a counting statement, as the indices m and n are the respective numbers of terms in the product, so, for instance,

$$a^2 \times a^3 = (a \times a) \times (a \times a \times a) = a^{2+3} = a^5.$$

In much the same way, we can make sense of the *second law of indices*, which says that $a^n/a^m = a^{n-m}$ through cancellation of common terms: for example,

$$\frac{a^5}{a^2} = \frac{a \times a \times a \times a \times a}{a \times a} = a^{5-2} = a^3.$$

Finally, the *third law of indices*, $(a^m)^n = a^{mn}$, is yet another counting statement: for example,

$$(a^2)^3 = (a \times a) \times (a \times a) \times (a \times a) = a^{2 \times 3} = a^6.$$

These rules show us the way to set up the definitions of negative and fractional powers, as that is done so that the three laws continue to be obeyed, a standard approach to generalization in mathematics. For instance, for any positive number a, by $a^{1/2}$ we mean $\sqrt{a}$, for then

$$a^{1/2} \times a^{1/2} = \sqrt{a} \times \sqrt{a} = a = a^1 = a^{1/2 + 1/2},$$

in accord with the first law. More generally, $a^{1/n}$ is defined to be the nth root of a. By a^{-1} we mean $1/a$, as this is consistent with the second law in situations such as

$$\frac{a}{a^2} = \frac{1}{a},$$

because subtracting the indices here leads to an index of $1 - 2 = -1$. The second law also requires that we define $a^0 = 1$ in order to be consistent with divisions such as $a^2/a^2 = 1$, as subtraction of the indices in this instance gives $2 - 2 = 0$.

The inverse operation of taking a power to a particular base is called the *logarithm* to that base. Each index law has a mirror image in this inverse process, which leads to the logarithms of products and quotients being equal to the sums and the differences of the logarithms, respectively. Since it is easier to add than to multiply, logarithms became the basis for complex calculations from the 17th century up until the time of the Moon landings.

There is a general expression describing the expansion of $(x + y)^n$ known as the *Binomial Theorem*. To find all the terms in the expansion, we need to sum the results of multiplying one term, either an x or a y, from each bracket. Since there are n brackets, and two terms in each bracket, this will lead to 2^n terms in all. However, there are only $n + 1$ different types of terms: if the variable x is chosen from a bracket on k occasions, where k could be any number ranging from 0 to n, then the alternative choice of y must be made on $n - k$ occasions, to give a term $x^k y^{n-k}$. We shall get one term $x^k y^{n-k}$ for each choice of k of the brackets $(x + y)$ from the list of n. The number of ways of choosing a set of k members from a set of n members is called a *binomial coefficient*, and this number crops up so often that it has its own notation: $\binom{n}{k}$. We will soon find the value of $\binom{n}{k}$ in terms of n and k, but in any event we have the following version of the expansion of $(x + y)^n$:

$$(x + y)^n = \binom{n}{0} x^0 y^n + \binom{n}{1} x y^{n-1} + \ldots + \binom{n}{k} x^k y^{n-k} + \ldots + \binom{n}{n} x^n y^0.$$

We can express this more compactly using *summation notation*: the Σ sign indicates that the following terms should be summed over the full range of values of k indicated above and below the Greek capital sigma:

$$(x+y)^n = \sum_{k=0}^{n} \binom{n}{k} x^k y^{n-k}. \tag{4}$$

There is one symmetry that often helps with binomial coefficients: whenever we choose a set of size k from a set of size n we simultaneously choose a set of size $n - k$, that being the complementary set of $n - k$ members of the n-set that were left behind when we chose the original k-set. It follows that we always have

$$\binom{n}{k} = \binom{n}{n-k}.$$

It is now simple to see that $\binom{n}{0} = \binom{n}{n} = 1$ and $\binom{n}{1} = \binom{n}{n-1} = n$. Applying this to the case $n = 3$ gives

$$(x+y)^3 = x^3 + 3x^2y + 3xy^2 + y^3,$$

and noting that there are six pairs that can be chosen from the set $\{1, 2, 3, 4\}$, which is to say $\binom{4}{2} = 6$, we obtain

$$(x+y)^4 = x^4 + 4x^3y + 6x^2y^2 + 4xy^3 + y^4.$$

We need, however, to know how to calculate the general binomial coefficient. The answer involves so-called *factorial* notation: we define $k!$, read as 'k factorial', as the product $k(k-1)(k-2)\ldots 2$, so, for example, $6! = 6 \times 5 \times 4 \times 3 \times 2 = 720$. The answer to our question is then

$$\binom{n}{k} = \frac{n!}{k!(n-k)!}, \tag{5}$$

although we need to adopt the convention that $0! = 1$ so that (5) gives the correct number in all circumstances. (This is similar to how we define $a^0 = 1$ in order that the index laws remain as general as possible.) To see that (5) is correct, we first ask for the number $P(n, k)$ of strings of length k that can be formed using k of the symbols $1, 2, \ldots, n$ (with no repeats allowed). We can choose the first number in n ways; for each such choice there remain $n-1$ ways to choose our second number, $n-2$ ways to choose the third number, and so on, there being $n-k+1$ ways of choosing the kth number. (Note that the first term in the product is n, not $n-1$, so the final term is $n-k+1$, and not $n-k$.) This provides a bridging expression:

$$P(n, k) = n(n-1)\ldots(n-k+1) = \frac{n!}{(n-k)!},$$

where the second equality is justified through cancellation of terms. Finally, each choice of a k-set, and there are $\binom{n}{k}$ of these, gives rise to $k!$ different strings of length k: for example, there are $4! = 24$ ways to arrange the four symbols a, b, c, d in a row. In this way, we obtain the result (5), for then

$$\binom{n}{k} \times k! = P(n, k) \quad \text{and so} \quad \binom{n}{k} = \frac{P(n, k)}{k!} = \frac{n!}{k!(n-k)!}.$$

For example,

$$\binom{7}{4} = \frac{7!}{4!3!} = \frac{7 \cdot 6 \cdot 5}{3 \cdot 2} = 7 \cdot 5 = 35.$$

Binomial coefficients arise constantly in enumeration problems and satisfy a host of identities that makes them very amenable to manipulation. Interesting facts are revealed by taking special values for the binomial terms x and y. To give one example, putting $x = y = 1$ in the Binomial Theorem (4) yields

$$(1+1)^n = 2^n = \sum_{k=0}^{n} \binom{n}{k}. \tag{6}$$

This has the following interpretation. We can in principle count all the possible subsets of a set of size n by adding up the numbers of all the subsets of size k from the least possible value of $k = 0$ (the *empty set*) to the maximum value of $k = n$ (the full set). This sum is of course just that of the right-hand side of (6), so, in other words, the total number of subsets of a set of size n is equal to 2^n.

The rules governing arithmetic

We close this chapter with some observations mainly concerning multiplication. There is one feature of its inverse operation, division, of which we need to be wary. We have already shown that $a \times 0 = 0$ is always true, and so it follows that 0 has no multiplicative inverse, which is to say there is no number a such that $a \times 0 = 1$. This is why our teachers impressed upon us the fact that 'you can't divide by 0', because the number $1/0$ does not exist.

In practical algebra, this represents a real pitfall. As we have already mentioned, the strength of algebra comes from manipulation of symbols in a fashion that is valid no matter which numbers are substituted for the symbols. However, when we *divide* by an algebraic expression, we must ensure that the expression does not represent 0, for if that were the case, we would be talking nonsense. We need to be mindful of this throughout our algebraic lives but, on the other hand, the vanishing denominator is the source of much that is wonderful in mathematics. Dividing into 0, however, is not a problem: for any $a \neq 0$, we have $0/a = 0 \times (1/a) = 0$.

Having established earlier that multiplication by 0 gives 0, we can now prove in a similar way that $(-1) \times (-1) = 1$. To prove this mysterious fact, we make use of the distributive law to argue as follows:

$$0 = (-1) \times 0 = (-1) \times (1 + (-1)) = (-1) \times 1 + (-1) \times (-1)$$

and so it follows that

$$0 = (-1) + (-1) \times (-1).$$

Adding 1 on the left to both sides of this equation gives us the desired result:

$$1 = (-1) \times (-1),$$

although we should note that in carrying out this final step, we use associativity of addition to bracket together $1 + (-1)$ so that the right-hand side becomes $0 + (-1) \times (-1) = (-1) \times (-1)$. This argument proves that 1 is the only value we may assign to $(-1) \times (-1)$ that remains consistent with the laws of algebra.

If a number x acts like the opposite of a number a, then it is its opposite, $-a$. To see this, suppose that $a + x = 0$; then adding $(-a)$ to both sides on the left and using associativity to bracket the first two terms on the left-hand side (LHS) together gives

$$((-a) + a) + x = (-a) + 0$$
$$\Rightarrow 0 + x = -a,$$
$$\Rightarrow x = -a,$$

where the symbol '$\Rightarrow$' is read 'implies': in general, $A \Rightarrow B$ means that if A is true, then B is true.

What we have here is the mathematical equivalent of the adage that if something looks like a duck and quacks like a duck, then it is a duck. Since the number x behaves like the opposite of a, it must indeed be $-a$. This fact is often expressed as saying that the inverse $-a$ of a is unique, meaning that no number other than $-a$ can be added to a to get the additive identity, 0. Similarly, $1/a$ is the only number that can be multiplied by a to get the multiplicative identity, 1. The reader may care to show by similar arguments that if $a + b = a$ then $b = 0$, while for non-zero a, if $ab = a$ then $b = 1$, and so the additive and the multiplicative identities of the integers are unique.

Using the fact that $a \times 0 = 0$ we may now, by expanding $a(b + (-b))$, see that $ab + a(-b) = 0$, and so, by the uniqueness of additive inverses, we conclude that $a(-b) = -(ab)$. In particular, it follows from this that the product of a positive and a negative number is negative. From this point we can derive that the product of two negatives is positive, a fact that we implicitly assumed when expanding $(a - b)^2$: two arbitrary negative numbers can be written as $-a$ and $-b$, where a and b are positive numbers. Using commutativity of multiplication, we now obtain

$$(-a) \times (-b) = (a \times (-1)) \times (b \times (-1))$$

$$= (a \times b) \times (-1) \times (-1) = a \times b.$$

The behaviour of the integers under multiplication may well be familiar. What we have done in this section, however, is to show that these rules are all algebraic consequences of the associative, commutative, and distributive laws along with the defining properties of 0 and 1 as the respective additive and multiplicative identity elements of the integers.

Chapter 3
Linear equations and inequalities

In this chapter we describe how to solve the simpler types of equations in one unknown, which are the linear equations, and simultaneous linear equations in two or more unknowns. We also introduce manipulation of inequalities, and the chapter closes with an example of finding the values of unknown quantities that satisfy particular equations while being constrained to lie within certain ranges.

Single equations with one unknown

Gilbert and Sullivan's model Major General boasted that he could solve equations both simple and quadratical. By 'simple' he meant ones of the form

$$ax + b = c,$$

where a, b, and c are given numbers and x is the unknown number to be found. Of course, there is no loss in assuming that the *coefficient* a of x is not zero. Any simple equation can be solved in two steps. We make x the subject of the expression by first subtracting b from both sides and then dividing by a, and in this way obtain

$$ax = c - b \quad \text{and so} \quad x = \frac{c-b}{a}.$$

This provides a formula by which to solve any simple equation. However, there is no need to rely on the formula, as we may carry out these steps in any particular case. For example, the equation $3x + 7 = 34$ is solved by saying $3x = 34 - 7 = 27$, and so $x = 27/3 = 9$.

The way we went about solving our equation has the seed of a fundamental idea. If we have *any* equation containing an unknown, x say, which appears only once, we may solve it by making x the subject of the formula represented by the equation.

But how do we do that? We unravel the process that embedded x into the equation in the first place: step by step, we reverse each operation, taking them in reverse order.

It is best to give an example to illustrate this general procedure. In this case, x is a number that is also the name of a film. I take x, subtract 4, multiply the resulting number by 2, then add 12, and finally divide the entire thing by 3, with the overall result being the number 7. What then is the name of the movie?

Beginning with the symbol x, the sequence of operations is as follows:

$$x \to x - 4 \to 2(x-4) \to 2(x-4) + 12 \to \frac{2(x-4) + 12}{3} = 7. \quad (7)$$

To extract the x from the equation in (7) we reverse each of the operations, doing this in the inverse order. Getting the order right is vital—after putting on your socks and shoes, it is important that you undress by *first* removing your shoes and then your socks. Always bear this *last-on first-off* principle in mind. The inverse process here yields a sequence of operations that we may represent symbolically as

$$\times 3 \to -12 \to \div 2 \to +4.$$

We carry out these four inverse operations in turn on both sides of our equation to yield a sequence of increasingly simple expressions:

$$2(x-4) + 12 = 3 \times 7 = 21$$
$$\Rightarrow 2(x-4) = 21 - 12 = 9$$
$$\Rightarrow x - 4 = \frac{9}{2}$$

and so

$$x = \frac{9}{2} + 4 = \frac{9+8}{2} = \frac{17}{2};$$

the film is Fellini's $8\frac{1}{2}$.

Our equation (7) is still a simple equation or, as they are more often called, a *linear* equation. A linear equation is one that can be written in the form $ax + b = c$, and (7) can be converted to that form for, starting from the $2(x - 4) = 9$ stage, we may expand the bracket by the distributive law to obtain $2x - 8 = 9$ and this is an equivalent equation, meaning that it is a consequence of and has the same solution as the original. The reason why the term *linear* is used stems from the general fact that the graph of all points (x, y) such that $y = ax + b$ is a straight line (Figure 3): the line *intercepts* the y-axis at the point $(0, b)$ and has *slope* (or *gradient*) a, meaning that for each unit moved in the positive x direction, we move a units up in the y direction.

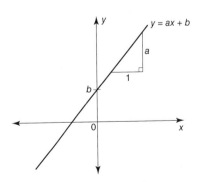

3. **Graph of a typical linear function $y = ax + b$.**

It may happen, however, that the unknown x appears more than once in our equation yet it may be possible, using just the laws of algebra, to reduce the equation to the standard linear form. As a more complicated example, let us take

$$\frac{x+38}{4-x} = 2.$$

We cannot treat the tangle on the left as the result of a sequence of operations applied to a single instance of x. The first simplification comes about, however, by multiplying both sides by the denominator, $4 - x$, as the LHS then becomes simply $x + 38$ and we have the following upon multiplying out the brackets:

$$x + 38 = 2(4 - x) = 8 - 2x.$$

We next add $2x$ to both sides in order to have only a single term in x. This gives

$$3x + 38 = 8$$
$$\Rightarrow 3x = 8 - 38 = -30$$
$$\Rightarrow x = \frac{-30}{3} = -10.$$

Simultaneous equations

At what temperature do the Celsius and Fahrenheit scales agree? That is to say, when do the two scales simultaneously give the same value? To answer this we need to know how the scales are calibrated. The Celsius scale is set at $0°$ at the temperature where water freezes, while $100°$ is the point where water boils (at sea level). These two temperatures are respectively marked at $32°$ and $212°$ on the Fahrenheit scale. If we let y denote temperature in Fahrenheit and x the value in Celsius, then the two are related by an equation of the form $y = ax + 32$ as when $x = 0$, $y = 32$. To find the value of the slope a we note that a rise of $(212 - 32)$ degrees Fahrenheit corresponds to an increase of $(100 - 0)$ degrees Celsius, so that:

$$a = \frac{212 - 32}{100 - 0} = \frac{180}{100} = \frac{9}{5}.$$

For example, if global warming led to an increase in the atmospheric temeperature by 2°C, this would represent an increase of $\frac{9}{5} \times 2 = \frac{18}{5} = 3 \cdot 6°$ Fahrenheit. However, a Fahrenheit *reading* (y) is related to a Celsius reading (x) by the linear equation $y = \frac{9}{5}x + 32$, so that an air temperature of 2°C corresponds to a temperature of $3 \cdot 6 + 32 = 35 \cdot 6°$ F. It is important not to confuse temperature changes with scale readings when discussing such topics!

Returning to our problem, we simultaneously require that $y = x$, so what we need to find is the point where the two lines that represent these equations cross (Figure 4).

It is natural to substitute $y = x$ into our conversion equation and solve

$$x = \frac{9}{5}x + 32 \Rightarrow 0 = \frac{4}{5}x + 32 \Rightarrow \frac{4}{5}x = -32;$$

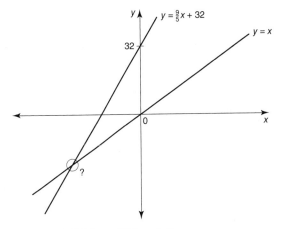

4. Temperature in Celsius and Fahrenheit.

multiplying both sides by the reciprocal, $\frac{5}{4}$, we extract the required value of x:

$$x = -32 \times \frac{5}{4} = -8 \times 5 = -40.$$

Therefore $-40°$ is the unique value where the Celsius and Fahrenheit scales agree.

This is an example of a problem that involves the solution of a pair of *simultaneous linear equations*, namely

$$y = \frac{9}{5}x + 32 \quad \text{and} \quad y = x.$$

Geometrically, the two equations represent a pair of straight lines and the problem is to find the point where the two lines meet. The points that lie along a particular line in the xy coordinate plane, often known as the Cartesian plane, are those points (x, y) that satisfy an equation of the form $ax + by = c$, where a, b, and c are fixed numbers that depend on and determine the line in question. Alternatively, we may rearrange this equation if we wish to make y the subject and we obtain the alternative slope–intercept form of the equation:

$$by = c - ax \quad \text{and so} \quad y = -\frac{a}{b}x + \frac{c}{b}.$$

We therefore turn to the general problem of finding the intersection point of two lines represented by general linear equations. Although only an example, the next problem is fully representative of the general situation, meaning that the way we solve it can be applied to any and all problems of this type. Our two equations are

$$-2x + 5y = 34,$$
$$3x + 4y = -5.$$

In our first example we had $y = x$ as our second equation, allowing us to substitute for y immediately. We could take that approach again, make y (or x) the subject of one of the equations, and substitute into the other accordingly. A different tack, though, is based on the idea that we could eliminate one of the variables simply by adding or subtracting the equations if the coefficients of one of the variables were opposites or were equal. In this example this is not the case, but we can get an equivalent pair of equations where this is so by multiplying the first equation throughout by 3 and the second by 2. Doing this, we find we may eliminate x by adding the two equations:

$$-6x + 15y = 102,$$
$$6x + 8y = -10,$$
$$23y = 92 \Rightarrow y = \frac{92}{23} = 4.$$

Now we substitute $y = 4$ into the first equation to give

$$5 \times 4 - 2x = 34 \quad \text{and so} \quad -2x = 34 - 20 = 14,$$
$$\text{whence } x = \frac{14}{-2} = -7.$$

Therefore the solution to our system of equations is $x = -7$, $y = 4$.

This *elimination* approach (adding and subtracting multiples of equations) is generally a quicker way to find the solution of the system, for it allows us to eliminate a variable using fewer steps than the substitution method. We shall take up an example with more than two unknowns in the final section of this chapter.

Substitution arises from first isolating a variable and then substituting accordingly, thereby eliminating that variable. The value of elimination is that you achieve the same thing while working solely with the coefficients. Matrices, which we introduce in Chapter 7, represent the vehicle for abstracting this process.

Inequalities

In this section we introduce the use and manipulation of inequality signs. Information about unknowns can come in the form of an equality, an equation if you will, or it can arise as a *constraint*, which can take the form of an inequality such as $2x - 1 \leq 5$. The signs $<$ and $\leq$ mean *less than* and *less than or equal to*, respectively, and of course the signs $>$ and $\geq$ stand for *greater than* and *greater than or equal to*, respectively. Each of these signs points to the smaller quantity in any statement in which it appears. Of course, the simple denial of equality, $a \neq b$, is an inequality, but the types of inequality that are most useful in mathematics are the directional inequalities as they often represent upper or lower bounds of quantities of interest and, with some important caveats that will now be explained, may be manipulated in a similar fashion to equalities.

We may be familiar with the meaning of an inequality such as $a > b$ when dealing with positive numbers, but we need to explain its meaning for any numbers, be they positive, negative, or zero. In line with our experience of positive numbers, we shall say quite generally that $a > b$ means that $a - b$ is a positive number. This definition orders the number line in the fashion that you probably take for granted: $b < a$ if b lies to the left of a on the line, so, for example, $-3 < 2$, $-8 < -\frac{1}{2}$, and $-1 < 0 < 1$ are all true statements.

Returning to our inequality $2x - 1 \leq 5$, we can make x the subject of the inequality in the same fashion as used when dealing with an equals sign: in this case we add 1 and then divide by 2 to simplify the constraint to $x \leq 3$.

There is, however, one significant complication that arises when dealing with inequalities that is a source of inconvenience and frequent error. Take an inequality such as $2 < 3$ and multiply both sides by a *negative* number, let us say -6. The left and right sides become respectively -12 and -18, and $-12 > -18$. And so we have another rule: when an inequality is multiplied (or divided) by a *negative* number, then the direction of the inequality is reversed. For example, let us simplify $4 - 3x \leq 13$. Subtracting 4 from both sides and then dividing through by -3 gives us

$$4 - 3x \leq 13 \Rightarrow -3x \leq 9 \Rightarrow x \geq -3.$$

This rule is a consequence of our definition. For instance, suppose that $a < b$ and let $c < 0$. Now, $a < b$ means that $b - a$ is positive, so that $c(b - a)$ is negative, which is to say that $cb - ca$ is negative, so its opposite, $ca - cb$ is positive, which tells us that $cb < ca$. In conclusion, if $a < b$ and $c < 0$ then $ca > cb$, and the direction of the inequality has indeed reversed.

Having cleared that up, it seems that we can continue confidently with our algebraic manipulations. When dealing with an inequality, however, we may wish to multiply both sides by an unknown, x say, but the resulting direction of the inequality depends on the sign of x. When this type of scenario arises we need to be patient and examine the cases that arise separately. We may sometimes, however, avoid multiplying by terms of unknown sign and thereby avoid splitting the problem into separate cases.

For example, suppose we wish to know for what values of x the following inequality holds:

$$\frac{x-1}{x+2} < -8.$$

If we now multiply throughout by $x + 2$, the inequality will change the direction if $x < -2$, but otherwise will not. Let us instead add 8 to both sides and place the LHS over a common denominator:

$$\frac{x-1}{x+2} + 8 < 0 \quad \text{and so} \quad \frac{(x-1) + 8(x+2)}{x+2} = \frac{9x+15}{x+2} < 0.$$

A common factor of 3 now appears in the numerator, which we extract and, remembering that $\frac{0}{3} = 0$, we may cancel to obtain

$$\frac{3(3x+5)}{x+2} < 0 \Leftrightarrow \frac{3x+5}{x+2} < 0. \qquad (8)$$

A quotient such as we have in (8) will be negative exactly when the numerator and denominator have different signs. A sign change may occur at a point where the term in question takes on the value 0, which is evidently at $x = -2$ for $x + 2$ and at $x = -\frac{5}{3}$ for $3x + 5$. It is helpful to chart the sign behaviour of each term on a number line.

From Figure 5, we see that the numerator and denominator have the same sign except between the values of -2 and $-\frac{5}{3}$, where $x + 2 > 0$ but $3x + 5 < 0$, and so that then is where our inequality holds: $-2 < x < -\frac{5}{3}$.

Inequalities are used throughout mathematics, often to find bounds of complicated functions in terms of simpler ones. Many useful inequalities stem from the simple observation that a square of a number is never negative. This follows from the fact that the product of two numbers with the same sign is positive. We shall give an instance of this, but first a word about square roots.

For any positive number a, by the *square root* of a we mean the unique positive number b such that $b^2 = a$. This is written as $b = \sqrt{a}$ so, for example, $\sqrt{25} = 5$. Of course, it is also the case that $(-5)^2 = 25$, so every positive number really has two square roots, $\pm\sqrt{a}$, although if we speak of *the* square root, implicitly we mean

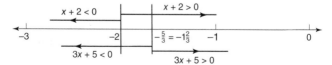

5. Signs of the expressions $x + 2$ and $3x + 5$.

the positive one. One useful property of square roots is that the square root of a product is equal to the product of the square roots, and, similarly, the square root of a quotient is the quotient of the square roots:

$$\sqrt{ab} = \sqrt{a} \cdot \sqrt{b}, \qquad \sqrt{\frac{a}{b}} = \frac{\sqrt{a}}{\sqrt{b}}.$$

That this is true is a consequence of the commutative law and the fact that two positive numbers are equal if and only if their squares are equal. To verify the first formula, then, we just need to check that the squares of both sides are the same: now, by the very definition of the square root, we have $\sqrt{ab}^2 = ab$, while

$$(\sqrt{a} \cdot \sqrt{b})^2 = \sqrt{a} \cdot \sqrt{b} \cdot \sqrt{a} \cdot \sqrt{b} = (\sqrt{a} \cdot \sqrt{a}) \cdot (\sqrt{b} \cdot \sqrt{b}) = ab,$$

and there is a similar tale for the quotient case. These rules are often used to simplify square roots of numbers that are not perfect squares. For example,

$$\sqrt{98} = \sqrt{49 \times 2} = \sqrt{49} \times \sqrt{2} = 7\sqrt{2}.$$

Since the sign $\sqrt{}$ indicates the non-negative root, it follows that if we begin with a *negative* number x, then $\sqrt{x^2}$ is not x but rather is $-x$: for example, $\sqrt{(-8)^2} = \sqrt{64} = 8 = -(-8)$. The function $\sqrt{x^2}$ has a special name: it is called the *absolute value* function and it has its own notation, $|x|$. Another way of thinking of $|x|$ is as the distance of x from 0 on the number line, an interpretation which generalizes nicely when we deal with complex numbers, which we shall meet in Chapter 5. Of course $|x|$ is always positive, except when $x = 0$, as $|0| = 0$.

The absolute value function is unpopular with just about everyone as, on the one hand, it seems so simple as to be hardly worth mentioning and, on the other hand, it misbehaves algebraically. Just like the square root, it does not behave linearly, meaning that just as it is *not* generally true that $\sqrt{a+b} = \sqrt{a} + \sqrt{b}$, nor is it true that $|a + b| = |a| + |b|$ (for example, if $a = 1$ and $b = -1$, the LHS is 0 while the right-hand side (RHS) is 2).

One rule, however, which is easily verified by looking at cases, is that $|ab| = |a| \cdot |b|$; so, for example, we may replace $|-2x|$ by $|-2| \cdot |x| = 2|x|$ in any calculation. In practice, you have to be patient and split a calculation involving absolute values into cases where the object between the absolute value signs is negative and where it is not. It is worth noting, however, that an expression such as $|x| \leq 3$ is equivalent to $-3 \leq x \leq 3$ and the latter is more amenable to algebraic manipulation. For example, let us simplify $|2x - 1| < 5$:

$$-5 < 2x - 1 < 5 \Leftrightarrow -4 < 2x < 6 \Leftrightarrow -2 < x < 3.$$

Returning to our search for inequalities based on squares, we introduce the *arithmetic mean* m of two numbers, a and b, which is what is normally referred to as their average: $m = (a + b)/2$. If we confine ourselves to non-negative numbers a and b, then the *geometric mean* g of a and b is defined as $g = \sqrt{ab}$. A square with sides of length g has the same area as an $a \times b$ rectangle.

For example, if $a = 4$ and $b = 9$ then $m = (4 + 9)/2 = 13/2 = 6\frac{1}{2}$, while $g = \sqrt{4 \times 9} = \sqrt{36} = 6$. If you experiment with a few examples of your own, you will discover that the arithmetic mean is never less than the geometric. And here is why:

$$0 \leq (\sqrt{a} - \sqrt{b})^2 = a - 2\sqrt{a}\sqrt{b} + b$$
$$\Rightarrow 2\sqrt{a}\sqrt{b} \leq a + b \Rightarrow \sqrt{ab} \leq \frac{a+b}{2},$$

which is just to say that $g \leq m$. Moreover, the initial inequality is a strict inequality except when $a = b$, in which case both means share this common value of a. In every other case, the arithmetic mean exceeds the geometric.

Lincoln Fair problem

As an example that brings together both the notion of inequalities and simultaneous equations, we close this chapter with the following problem that dates back to the 16th century, if not earlier. Twenty people pay twenty pence to visit the Lincoln Fair. If each man pays threepence, each woman tuppence, and each child a halfpenny, how many men, M, women, W, and children, C, went to the fair?

We can take the information given to extract two equations in *three* unknowns—the first counts people while the second counts pennies:

$$\begin{aligned} M + W + C &= 20, \\ 3M + 2W + \tfrac{1}{2}C &= 20. \end{aligned} \tag{9}$$

Having more unknowns than equations generally means you do not have enough information to solve the problem, or, to be more precise, there is more than one solution. As we have already mentioned, the equation $ax + by = c$ represents a *line* in the xy plane and, in an analogous way, an equation of the form $ax + by + cz = d$ represents a *plane* in 3D x, y, z coordinates. Two such planes generally intersect in a line, and so any one of the infinite number of points along that line will have coordinates (x, y, z) that simultaneously satisfy each of the equations of the two planes. Undaunted, we see how far our elimination technique can take us in this problem. Multiplying the first equation by 3 and then subtracting the second equation will at least eliminate the men, giving us

$$W + \frac{5}{2}C = 40 \Rightarrow W = 40 - \frac{5}{2}C. \tag{10}$$

We can now also express M in terms of C by using our first equation:

$$M = 20 - W - C = 20 - \left(40 - \frac{5}{2}C\right) - C;$$

When we remove the brackets in this last expression, we must remember to subtract every term inside the brackets (and not just the first one); subtracting the term $-\frac{5}{2}C$ of course gives $+\frac{5}{2}C$. Continuing and rearranging the order of the terms, we infer that

$$M = \frac{5}{2}C - C + 20 - 40 = \frac{3}{2}C - 20. \tag{11}$$

We have managed to express M and W in terms of C. If we knew no more than the pair of equations (9), we could go no further: we could choose *any* number for C and determine W and M from C using (10) and (11), and the trio of numbers (C, W, M) that resulted would be a solution to (9); and, what is more, every solution to (9) would arise in this way. We do, however, know more, although the additional information not captured by our simultaneous equations (9) comes in the form of inequalities.

You cannot have fractional or negative people, and so, assuming that there was at least one person of each of the three types, we have the three inequalities $C, W, M \geq 1$. There is another more subtle point that is revealed by looking at the second of our two equations in (9): in order that the left-hand side adds up to the whole number 20, the number of children must be even. Hence we may write $C = 2A$, say, where A is itself a positive integer. Using this and (10) and (11) allows us to write our inequality for W as follows:

$$W = 40 - 5A \geq 1 \quad \text{and so} \quad 40 \geq 1 + 5A$$

$$\Rightarrow 5A \leq 39 \quad \text{and so} \quad A \leq \frac{39}{5} = 7\frac{4}{5}$$

$$\therefore A \leq 7 \text{ as } A \text{ is a whole number;}$$

Similarly, $M = 3A - 20 \geq 1 \Rightarrow 3A \geq 21$ and so $A \geq 7$.

Hence we simultaneously have $A \leq 7$ and $A \geq 7$, so that $A = 7$ and therefore $C = 2A = 14$. There were $C = 14$ children, and so $M = 3A - 20 = 21 - 20 = 1$, and $W = 40 - 5A = 40 - 35 = 5$.

Therefore one man, five women, and 14 children went to the Lincoln Fair.

Whenever any system of equations is solved, we may verify the solution by substituting: in this example, putting $C = 14$, $W = 5$, and $M = 1$ into each of our equations in (9) shows that our solution does indeed work. As our reasoning has proved, this is the unique solution to the Lincoln Fair problem provided that we assume there was at least one man, one woman, and one child. Some of the unknowns could, however, be permitted to take on the value 0 without reducing the problem to total nonsense. This would weaken the constraints to $C, W, M \geq 0$. If you rework the problem, you find that a second value, $A = 8$, is now also feasible. Putting $A = 8$ then gives a second solution: $(C, W, M) = (16, 0, 4)$: 16 children, no women, and four men does, strictly speaking, still work.

There is an entire realm of mathematics, known as *linear programming*, that is dedicated to solving huge systems of linear equations subject to constraints on the values of the solutions. Our Lincoln Fair problem is a very simple historical instance of this kind of question. This mathematics underpins much of the logistics of modern society, governing operations such as railway and airline schedules while allowing businesses to meet customer demand by running inventories with much less stock stored in warehouses than was required in times past, which represents an enormous and ongoing cost saving. The algebraic ideas of elimination and constraint satisfaction underlie all this. Contemporary society could not function without them operating, behind the scenes, invisibly and flawlessly on behalf of everyone.

Chapter 4
Quadratic equations

A *quadratic equation* is one involving a squared term and which, when all terms have been moved to the left-hand side, takes on the form $ax^2 + bx + c = 0$. Quadratic equations are as old as mathematics itself, with examples appearing in ancient Babylonian tablets from over 4000 years ago. What is more, although lacking modern notation, these ancient mathematicians seemed to be well aware of the principles involved in solving them. Quadratic expressions are central to mathematics, and quadratic approximations of all kinds are tremendously useful in describing processes that are changing in direction from moment to moment.

The way to tackle quadratic equations is normally taught in three stages. The first looks at simple examples and tries to factorize the quadratic expression into two linear factors, which then allows you to write down the two solutions (for normally there are two). This seems to be guesswork, trial and error if you like, and so not satisfactory. Next, a mathematical device is introduced called *completing the square*, which allows solution of any particular quadratic. This is the general method of the ancients, although the key step may seem at first to be a little unnatural. Finally, completing the square is applied to the general equation to derive the famous quadratic formula that allows us to plug the three coefficients into the associated expression, which then provides the solutions. This formula often represents the first sophisticated piece of algebra that a student meets.

Although the quadratic formula solves every quadratic, there is much to be learnt from the other approaches apart from just methods to solve the equations, as will now be explained.

Factorization and completing the square

Let's begin with the equation $x^2 - 5x + 6 = 0$. The idea is to find a factorization of the expression as $x^2 - 5x + 6 = (x - r)(x - s)$, for then the two solutions of our equation will be the numbers r and s, known as the *roots* of the equation. This is because, when we substitute $x = r$ into this factorized expression it returns $(r - r)(r - s) = 0(r - s) = 0$ and, similarly, substituting $x = s$ into this product also gives the desired output of 0, while no other numbers apart from r and s can do this. But how are we to find r and s?

The next step represents an important idea. We expand the expression $(x - r)(x - s)$ to give it the same form as the original quadratic expression and then choose r and s to match the corresponding coefficients. Now,

$$(x - r)(x - s) = x^2 - rx - sx + rs = x^2 - (r + s)x + rs.$$

We equate the two expressions to hand:

$$x^2 - 5x + 6 = x^2 - (r + s)x + rs,$$

so we need to find r and s so that $r + s = 5$ and $rs = 6$. A little thought will lead to the result $r = 2$, $s = 3$, which is to say that $x^2 - 5x + 6 = (x - 2)(x - 3)$, so the solutions of our original equation are $x = 2$ and $x = 3$. This works—substituting either the number 2 or the number 3 into $x^2 - 5x + 6$ does indeed give zero.

The idea here is to take the quadratic expression $x^2 + bx + c$ and write it in the form $(x - r)(x - s)$. Expanding the latter expression and matching coefficients then gives a pair of simultaneous equations in the two unknowns r and s, one of which, $r + s = -b$ is linear, but the other, $rs = c$, is not. We might try to solve these equations by substituting $s = -r - b$ into the second equation, $rs = c$, but that leads us in a circle back to the original quadratic equation, so this approach has serious shortcomings.

Clearly we need a better method, but before moving on it is worth noting that this sum-and-product formulation is equivalent to the original equation, and indeed this was the way these questions were often posed in classical texts, as they arise through the problem of determining the dimensions of a rectangle given its area and perimeter. With this kind of formulation, the problem was bound to have positive solutions, which were the only quantities that the ancients recognized as true numbers. For example, let us suppose that the perimeter of the rectangle is given as 28 units and the area as 48 units^2. Let r and s denote the unknown dimensions of the rectangle, giving us the pair of equations

$$2(r + s) = 28 \quad \text{and so } r + s = 14, \quad \text{and } rs = 48.$$

From the linear equation we infer that $s = 14 - r$, and substituting accordingly into the area equation yields

$$r(14 - r) = 48 \Rightarrow -r^2 + 14r = 48 \Rightarrow r^2 - 14r = -48.$$

Instead of guessing and checking, we may *complete the square* to solve this quadratic. Compare the LHS, $r^2 - 14r$, to the general expression $r^2 - 2ry + y^2 = (r - y)^2$. The first two terms have the right form but the final term, y^2, is missing. We may, however, *force* our expression to have the desired form by adding the absent term, although in order to retain a correct equation we add that term to the RHS as well to keep it balanced.

Specifically, here we will need to match $-14r$ with $-2ry$ so that $-14 = -2y$; hence $y = 7$ and its square is $y^2 = 7^2 = 49$. This is the magic ingredient that when added to both sides allows us to take square roots:

$$r^2 - 14r = -48 \quad \text{and so } r^2 - 14r + 49 = -48 + 49 = 1,$$
$$\text{whence } (r-7)^2 = 1 \quad \text{and so } r - 7 = \pm 1,$$

which is to say $r = 7 \pm 1$.

Therefore we get that either $r = 7 - 1 = 6$, whereupon $s = 14 - r = 14 - 6 = 8$, or $r = 7 + 1 = 8$ and $s = 14 - r = 14 - 8 = 6$. Either way we are led to what is essentially the unique solution, that being a 6×8 rectangle.

The overall approach to completion of the square is as follows. Given any quadratic equation $ax^2 + bx + c = 0$ with $a \neq 0$, we first divide through by a to get a *monic* equation—one in which the coefficient of the highest power, x^2, is 1. This is an equivalent equation, so it follows that if we can solve any monic quadratic equation, we can solve them all, and so there is no loss of generality in confining ourselves to equations that can be written in the form $x^2 + bx = c$. We can complete the square on the LHS by adding the square of $b/2$ (and not of b) because

$$x^2 + bx + \left(\frac{b}{2}\right)^2 = \left(x + \frac{b}{2}\right)^2,$$

as the cross-term that emerges when this square is expanded is $2 \cdot (b/2)x = bx$, as it needs to be. Hence, by adding this particular square to both sides, we have a new form of the equation in which the LHS is an explicit square. We now take square roots (both positive and negative) and make x the subject of the expression to finally get to the roots.

It should be appreciated that there are other interesting questions that arise involving quadratic expressions, and completion of the square allows us to answer them. We give two examples.

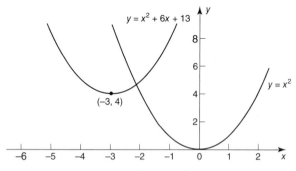

6. Comparing the graphs of $y = x^2$ and $y = x^2 + 6x + 13$.

The curve described by the equation $y = x^2$ is the standard upward-opening parabola that has the y-axis as its axis of symmetry, and its vertex is the origin $(0, 0)$. How does the graph of the equation $y = x^2 + 6x + 13$ compare with that of $y = x^2$? (See Figure 6.)

By completing the square, we can show that the two curves are identical; the difference is just that the second graph has been translated to another position, which we can describe precisely. Since $\frac{6}{2} = 3$ and $3^2 = 9$, we may complete the square as follows:

$$y = x^2 + 6x + 13 = (x^2 + 6x + 9) + 4 = (x + 3)^2 + 4.$$

Now the minimum of $y = (x + 3)^2$ is 0, which occurs when $x = -3$: overall, the graph of $y = (x + 3)^2$ is that of $y = x^2$, only moved 3 units to the left. To obtain the graph of $y = (x + 3)^2 + 4$, we now move the graph of $y = (x + 3)^2$ by 4 units in the positive y-direction. In summary, the graph of $y = x^2 + 6x + 13$ is realized by shifting the graph of $y = x^2$ by 3 units to the left and 4 units up. In particular, the turning point of this parabola rests at $(-3, 4)$.

Our second problem is to find the number x that most exceeds its own square. This is to say, we wish to find the value of x that

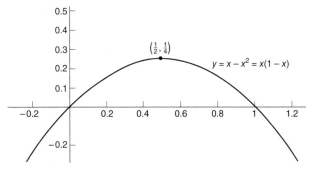

7. Graph of $y = x - x^2$.

maximizes $x - x^2$ or, what is the same, that minimizes its negative, $x^2 - x$ (Figure 7). Completing the square for this expression requires us to add (and then subtract) $\left(-\frac{1}{2}\right)^2 = \frac{1}{4}$. We then have

$$x^2 - x = \left(x^2 - x + \frac{1}{4}\right) - \frac{1}{4} = \left(x - \frac{1}{2}\right)^2 - \frac{1}{4}.$$

We minimize this quantity by making the square as small as possible, which is to say 0^2, and this is done in this case by putting $x = \frac{1}{2}$. That then is our answer: $\frac{1}{2}$ is the number that most exceeds its own square. We may note that $\frac{1}{2} - \left(\frac{1}{2}\right)^2 = \frac{1}{2} - \frac{1}{4} = \frac{1}{4}$ and any other number exceeds its square by a lesser amount—indeed, for any number x outside the interval $0 \leq x \leq 1$, x is actually less than x^2.

As a final example in this section, let us solve

$$x + \sqrt{x+1} - 11.$$

A natural impulse is to rid ourselves of the awkward square root through squaring, but doing that immediately would not work, as the root would come back to haunt us in the cross-product that arises in the square of the LHS. Instead, we focus our attack on the root by first making it the subject of the equation. Only then do we square:

$$\sqrt{x+1} = 11 - x \Rightarrow x + 1 = (11-x)^2 = (x-11)^2 = x^2 - 22x + 121;$$

$$\text{hence } x^2 - 23x + 120 = 0 \quad \text{and so } (x-8)(x-15) = 0,$$

which gives the two solutions $x = 8$ and $x = 15$. However, if we test these solutions, only one of them works in the original equation, $x + \sqrt{x+1} = 11$:

$$8 + \sqrt{8+1} = 8 + 3 = 11; \quad 15 + \sqrt{15+1} = 15 + 4 = 19.$$

What has gone wrong here is that the squaring of both sides of the equation has introduced extraneous solutions, for the implication sign cannot be reversed: in general, $a = b \Rightarrow a^2 = b^2$, but $a^2 = b^2 \not\Rightarrow a = b$, but only that $a = \pm b$. The extraneous solution, 15, represents the solution to the equation $x - \sqrt{(x+1)} = 11$.

The general point that needs to be appreciated is that if we carry out an operation on both sides of an equation that cannot be reversed, then the new equation is *not* equivalent to the original. Any solution of the original equation will nonetheless be contained in the solution set of the new equation. Therefore, having solved the new equation, we need to test the solutions on the original as not all of them will necessarily apply.

The quadratic formula

Applying the method of completing the square to the general quadratic equation yields the general quadratic formula, which solves the entire class of problems in one fell swoop. Quite naturally, this chapter culminates in this derivation. The slickest way to produce the formula comes through building up the RHS to the required expression as directly as possible. This begins by writing the equation as $ax^2 + bx = -c$, multiplying throughout by $4a$, and then adding b^2 to both sides to complete the square on the left. In detail, we have

$$ax^2 + bx = -c$$

$$\Rightarrow 4a^2x^2 + 4abx = -4ac$$

$$\Rightarrow 4a^2x^2 + 4abx + b^2 = b^2 - 4ac$$

$$\Rightarrow (2ax + b)^2 = b^2 - 4ac$$

$$\Rightarrow 2ax + b = \pm\sqrt{b^2 - 4ac};$$

$$\therefore x = \frac{-b \pm \sqrt{b^2 - 4ac}}{2a}.$$

This derivation avoids any fractional terms until the very last line, although the argument is based upon knowing where we are headed.

It is worth studying another derivation that introduces a more general approach that may apply to other problems. The following method does not require an inspired observation as to how to manufacture squares when we are not given them in the first place, but rather the completion emerges from the algebra itself. Again we begin with the general quadratic equation, $ax^2 + bx + c = 0$ with $a \neq 0$. We make the substitution $x = y + t$, where t is a constant to be specified. When we make this substitution, the coefficient of y will involve t and the other coefficients. If we then choose t so that the term in y vanishes (because its coefficient is 0), we will have an equation of the form $y^2 = k$ for some constant k, from which point we may solve through taking square roots. With this in mind, we now put $x = y + t$. Upon rearranging the order of the terms that arise through the expansion, we obtain:

$$a(y+t)^2 + b(y+t) + c = ay^2 + (2at + b)y + (at^2 + bt + c) = 0. \quad (12)$$

We now choose t so that the multiplier of y is 0, which is to say that the special value of t we seek comes about by putting

$$2at + b = 0 \Rightarrow t = -\frac{b}{2a}. \quad (13)$$

The term in (12) involving y^2 is ay^2, while by making use of (13) we get the following for our constant term k:

$$k = at^2 + bt + c = \frac{b^2}{4a} - \frac{b^2}{2a} + c = \frac{b^2 - 2b^2 + 4ac}{4a} = \frac{4ac - b^2}{4a}.$$

We now have a clear path to the general formula, for we have reduced our original equation to $ay^2 + k = 0$. Writing this as $y^2 = -k/a$, we have

$$y^2 = \frac{b^2 - 4ac}{4a^2}.$$

Rewriting y as $x - t = x + b/2a$ and then taking square roots, the previous equation now gives

$$y = x + \frac{b}{2a} = \frac{\pm\sqrt{b^2 - 4ac}}{2a};$$

$$\therefore x = \frac{-b \pm \sqrt{b^2 - 4ac}}{2a}. \tag{14}$$

The Indian mathematician Brahmagupta (AD 597–668) seems to have been the first to explicitly describe the general quadratic formula, albeit in words, in his treatise *Brāhmasphuṭasiddhānta*, published in AD 628. Earlier still, Diophantus adopted a recognizably algebraic approach to such problems in the 3rd century AD. (The solutions of Euclid, some six centuries earlier, represented the roots geometrically.) The formula was, however, first stated in the modern form in 1594 by the Flemish mathematician Simon Stevin (1548–1620), who also brought decimal calculation into common usage in Europe.

As an example that leads to a quadratic the roots of which are not integers or fractions, we shall find the dimensions of the Golden Rectangle, which is defined as the rectangle that has the property that when the largest possible square is removed from the rectangle, the smaller rectangle that remains is similar to the original, meaning that both rectangles have the exact same shape, the only difference being one of scale (Figure 8).

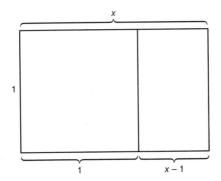

8. Golden Rectangle.

But what is that scale? If we take the shorter side of the parent rectangle as 1 unit and the long side as x, we have from the given similarity condition that

$$\frac{x}{1} = \frac{1}{x-1} \Rightarrow x(x-1) = 1^2 \Rightarrow x^2 - x - 1 = 0. \tag{15}$$

Applying the quadratic formula (14) to the equation of (15) requires that we put $a = 1$, $b = c = -1$ and so we obtain

$$x = \frac{-(-1) \pm \sqrt{(-1)^2 - 4(1)(-1)}}{2(1)} = \frac{1 \pm \sqrt{1+4}}{2};$$

and since the positive solution is the one we want, we find that the Golden Ratio, often denoted by the Greek letter ϕ, is given by

$$\phi = \frac{1 + \sqrt{5}}{2}. \tag{16}$$

The Golden Ratio arises persistently throughout mathematics, especially in the context of self-similarity problems.

We end this section by examining the number of solutions a quadratic equation may have. If we graph the function $y = ax^2 + bx + c$, the solutions of the corresponding quadratic equation $y = 0$ tell us where that graph meets the x-axis. Of course, the graph may not cross the axis at all.

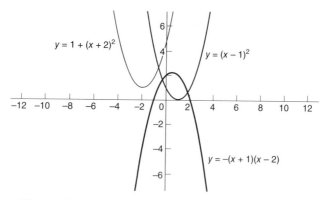

9. Three quadratic graphs exhibiting 0, 1, and 2 roots, respectively.

It all depends on the *discriminant* $\Delta = b^2 - 4ac$, which is the term that lies under the square root sign in the quadratic formula (14); in the monic case, $\Delta = (r - s)^2$, the square of the difference of the roots. The quadratic has two solutions if $\Delta > 0$ but no solutions if $\Delta < 0$, as there is no square root of a negative number. If $\Delta = 0$, however, there is a unique solution, that being $x = -b/2a$, in which case the corresponding graph just touches the x-axis at this point, with the axis being tangent to the curve (see Figure 9). We close with a problem that showcases this transitional case.

A man, running at constant speed v, tries to catch an old-fashioned London bus (of the kind that you could leap on while in motion) (Figure 10). When he gets within a distance d of the door, the bus, which has been stationary, begins moving away from him with constant acceleration a. What is the maximum value of d that will allow our man to catch his bus?

Let us take the origin, O, from which we measure the position of both man and bus to be the man's initial position as the bus begins to move. At time t he will have travelled a distance vt in the positive x-direction. The *speed* of the bus at time t, however, is at. Since it started from rest and the acceleration is constant, the

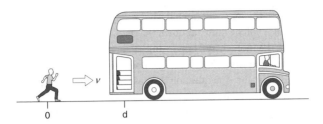

10. Man running for a bus.

average speed of the bus in the time interval from time 0 up to time t is $(at - 0)/2 = \tfrac{1}{2}at$. Hence the distance of the door of the bus from the origin at time t is $d + \left(\tfrac{1}{2}at\right)t = d + \tfrac{1}{2}at^2$.

Now, the runner will be at the same position as the door at the times when these two expressions for the position of man and bus agree, which is to say when the time t satisfies

$$vt = d + \frac{1}{2}at^2 \Rightarrow 2vt = 2d + at^2$$
$$\Rightarrow at^2 - 2vt + 2d = 0. \tag{17}$$

We could now find these times by solving the equation (17), which is a quadratic equation in the variable t of time. However, that was not the question. Take a moment to think what is physically possible here.

Suppose there are two solutions to (17). The earlier of the two times would represent the moment when the man catches up with the bus. If he decided not to get on the bus at this point, he would overrun the door. However, since he is running at constant velocity and the bus is accelerating, eventually the bus would catch up and pass him by. The second solution is the later point when the bus door would pass him, offering him his last chance to hop on board. If, on the other hand, the values of d and a are too great, he never catches the bus and (17) has no solutions. These two scenarios correspond to $\Delta > 0$ and $\Delta < 0$, respectively. If the values of a and

d are small, the man will catch his bus easily and the second coincidence will be quite far down the road. As we increase a and/or d, the moments of opportunity to board draw closer to one another, and at the point where $\Delta = 0$ they merge into a single precious moment. The question then asks us to find, for a given value of a, what is the value of d where there is just one solution to (17). To determine this value, we just need to set $\Delta = 0$. First, let's find Δ for our equation (17):

$$\Delta = (-2v)^2 - 4a(2d) = 4v^2 - 8ad.$$

Putting $\Delta = 0$ now gives the critical value of d:

$$8ad = 4v^2 \quad \text{and so} \quad d = \frac{4v^2}{8a} = \frac{v^2}{2a}.$$

The would-be passenger will catch his bus as along as $d \leq v^2/2a$. For example, if the man is running at 6 m/s and the bus is accelerating at 1 m/s² then the critical value is $d = 6^2/2(1) = 18$, so if the initial gap between him and the door is more than 18 m, then the bus gets away.

Having solved linear and quadratic equations, our next step would seem to be the solution of equations involving cubes and higher powers. Prior to that, however, we shall look at the algebra of polynomials and how it compares to that which governs the number system in order to see what may be learnt in general about equations involving powers higher than 2.

Chapter 5
The algebra of polynomials and cubic equations

General polynomials

We prepare the ground for solving the cubic by first investigating the nature of *polynomials*, which are expressions of the form

$$p(x) = a_0 + a_1 x + a_2 x^2 + \ldots + a_n x^n;$$

we call $p(x)$ a polynomial of *degree n*. The number a_i is called the *coefficient* of x^i, a_0 is the *constant term* of $p(x)$, and a_n is called the *leading coefficient*. For example, $p(x) = -9 - 5x + x^2 - 2x^3$ is a third-degree or *cubic* polynomial. We may sometimes wish to find the solutions x of the equation $p(x) = 0$, but we will also consider polynomials as objects in their own right, in which case the role of the symbol x is just that of a placeholder—if another letter such as y or t were used, we would have essentially the same polynomial. Since the polynomial $p(x)$ is determined by the list of coefficients $(a_0, a_1, a_2, \ldots, a_n)$, we could *define* the polynomial as being this list and dispense with the symbol x entirely. In practice, however, $p(x)$ is often the rule for a function and so we consider x to be a variable, which is to say that x may be any number $x = a$, and $p(a)$ is the value of its image when the polynomial is evaluated at $x = a$. For our cubic polynomial we have, for example,

$$p(-2) = -9 - 5(-2) + (-2)^2 - 2(-2)^3 = -9 + 10 + 4 + 16 = 21.$$

In general, a *function* can be thought of as any rule whatsoever that takes as input a given number and outputs another (generally different) number—this idea is fundamental to modern mathematics. The importance of polynomials as functions is twofold. First, the outputs of polynomial functions can be explicitly calculated, and second, polynomials are very versatile, for they may be used to approximate many important functions whose output values often can be calculated only imperfectly, such as the trigonometric, exponential, and logarithm functions.

The underlying algebra of polynomials mirrors that of the integers both in obvious and in more subtle ways. We may add, subtract, and multiply polynomials, and the laws of algebra, namely associativity, commutativity, and the distribution law of addition over multiplication, all hold. This stems from the fact that we may regard the variable x as representing an arbitrary number, so these expressions may be added and multiplied in the usual way and the above laws are bound to persist because we have defined the addition and multiplication operations with this in view. For example,

$$(1 + x^2)(1 - x - x^5) - 2x^3 + x^4$$
$$= (1 - x - x^5) + (x^2 - x^3 - x^7) - 2x^3 + x^4$$
$$= 1 - x + x^2 - 3x^3 + x^4 - x^5 - x^7.$$

Note that both terms in the first bracket need to be multiplied by each of the terms in the second, not forgetting to attribute the correct sign to each product. This yields a contribution of $2 \times 3 = 6$ terms in all. We then simplify by collecting up like powers. The degree of the final polynomial is then the sum of the degrees of the highest powers in the terms of the product, which in this case is $2 + 5 = 7$.

Polynomial division is especially interesting. The situation regarding integers a and b, allowing also for b to be negative, is that there are unique integers q and r such that $a = bq + r$, with the remainder, r, satisfying $0 \leq r \leq |b| - 1$. There is an analogue of this equation for polynomials where the measure of the size of the polynomial is its degree. In particular, if $a(x)$ and $b(x)$ are polynomials of degrees n and m, respectively, with $m \leq n$, then we may find an expression $a(x) = q(x)b(x) + r(x)$, where the degree of the remainder polynomial, $r(x)$, is less than m.

The important case for us here is the simple one where $b(x)$ is a linear monic polynomial, i.e. $b(x) = x - c$, and then the remainder polynomial has degree 0, which is to say $r(x)$ is simply a constant. For example, let us take $a(x) = 1 + x - 3x^3 + x^4$ and $b(x) = x - 4$; we need to find the quotient polynomial, $q(x) = c_0 + c_1x + c_2x^2 + c_3x^3$, and the remainder, r. Our polynomial equation is then

$$1 + x - 3x^3 + x^4 = (c_0 + c_1x + c_2x^2 + c_3x^3)(x - 4) + r. \qquad (18)$$

Starting from the highest power in the expansion, x^4, we immediately get $c_3 = 1$. Equating terms in x^3 then gives $c_2 - 4c_3 = c_2 - 4 = -3$ so that $c_2 = -3 + 4 = 1$. Comparing coefficients of x^2 then gives $c_1 - 4c_2 = c_1 - 4 = 0$, whence $c_1 = 4$; next, looking at the term in x, we see that $c_0 - 4c_1 = c_0 - 16 = 1$, and so $c_0 = 17$. Finally, for the remainder term we have $r - 4c_0 = r - 68 = 1$, so that $r = 69$. In conclusion, we have

$$1 + x - 3x^3 + x^4 = (17 + 4x + x^2 + x^3)(x - 4) + 69.$$

What is important here, however, is that this form of calculation may always be carried out in the fashion of this example: comparing the highest powers gives the coefficient c_{n-1} of x^{n-1} in $q(x)$, and then, working with the powers in descending order, we obtain a linear equation for each coefficient c_i in terms of a_i and c_{i+1}, which is now known, so each of the numbers $c_{n-1}, c_{n-2}, \ldots, c_0, r$ may be found in turn.

Long division of one polynomial by another is important in practical calculations in integral calculus, but here it is the theoretical fact that the equation $a = qb + r$ for polynomials is soluble that allows us to deduce two important theorems, the second of which is a corollary of the first, these being the *Remainder* and *Factor Theorems*.

When we divide $a(x)$ by $x - c$, we can find r immediately without needing to know the quotient polynomial $q(x)$, for consider the equation

$$a(x) = q(x)(x - c) + r;$$

if we substitute $x = c$ into this equation, we obtain

$$a(c) = q(c)(c - c) + r = 0 + r = r.$$

This is the Remainder Theorem: the remainder when we divide a polynomial $a(x)$ by the term $x - c$ is equal to $a(c)$. For example, the remainder r in (18) is given by $a(4)$, so that

$$r = a(4) = 1 + 4 - 3(4^3) + 4^4 = 5 + 4^3(-3 + 4) = 5 + 4^3 = 5 + 64 = 69,$$

which agrees with the value we found from the full division calculation.

If the remainder $r = 0$, we say that $x - c$ is a *factor* of $a(x)$, for then we have $a(x) = (x - c)q(x)$, and so $a(c)$ is zero and c is a *root* of the polynomial $a(x)$. Conversely, if c is a root of $a(x)$, which is to say that $a(c) = 0$, then by the Remainder Theorem the remainder when $a(x)$ is divided by $x - c$ is 0, which gives $a(x) = (x - c)q(x)$. In summary, we have the Factor Theorem: c is a root of the polynomial $a(x)$ if and only if $a(x)$ has a factorization of the form $a(x) = (x - c)q(x)$.

If d were another root of $a(x)$ then $0 = a(d) = (d - c)q(d)$, so that $q(d) = 0$, and by the Factor Theorem we have $q(x) = (x - d)p(x)$, say, where $p(x)$ is a polynomial of degree $n - 2$. Since each root t of $a(x)$ introduces a new factor, $(x - t)$, into the factorization of $a(x)$ and the degree of the associated quotient polynomial decreases by 1 each time, we infer that the number of distinct roots of an nth-degree polynomial $a(x)$ is at most n. If $a(x)$ has n roots, $c_1, c_2, \ldots, c_n$, then $a(x)$ will have a factorization of the form

$$a(x) = A(x - c_1)(x - c_2)\ldots(x - c_n) \qquad (19)$$

for some constant A, which evidently is then the leading coefficient a_n of $a(x)$.

Complex numbers

The totality of all numbers represented by points on the number line is referred to as the set of *real numbers* and is denoted by $\mathbb{R}$. We can think of the positive real numbers as all those that can be represented in decimal form. This is a much larger collection than the rationals, $\mathbb{Q}$, as the decimal expansion of any fraction m/n, is always recurring, with the recurring block no longer than $n - 1$: for example, $2/7 = 0.\overline{285714}$ shows a recurring block of length $7 - 1 = 6$. Decimals that never fall into a recurring pattern therefore represent real numbers that are not rational: for example $0.10110111011110\ldots$ is an irrational number, as is $0.12345678910111213\ldots$. The decimal expansion of $\pi = 3.14159\ldots$ is seemingly random in nature—π is certainly not a rational number, although proving that fact is difficult.

If we identify the notion of a number with that of a signed distance from 0, measured either to the left (negative) or right (positive) along the number line, then the real numbers are all the numbers to be found. This collection does, however, have one serious shortcoming: since squares are never negative, it follows that negative numbers have no square roots. In particular, there is no real number a such that $a^2 = -1$.

Mathematicians decided this would not do, as the lack of such a number acted as a roadblock to the kind of free algebraic movement required to solve certain equations. To circumvent the obstacle, the *imaginary unit i* was introduced, which was endowed with the property that $i^2 = -1$. This idea may also have been motivated via the number line itself, beginning with the observation that multiplying the members of the number line by -1 has the effect of rotating the line through $180°$ about 0, with 1 going to -1, 3 being mapped to to -3, and so forth. Since $i^2 = -1$, it seems that *two* multiplications by i are needed in order to have the same effect, which suggests that multiplication by i will rotate the number line through $90°$ about the origin. This right-angled rotation creates another copy of the number line at right angles to the original and delivers a plane of numbers, known as the *Argand plane*, named after Jean-Robert Argand (1768–1822). The point in this plane with coordinates (a, b) then corresponds to what is known as the *complex number* $z = a + ib$, where a and b are ordinary real numbers, known respectively as the *real* and the *imaginary* parts of the complex number z. The intention is that by adjoining the new number i to $\mathbb{R}$, our expanded collection of complex numbers, denoted by $\mathbb{C}$, should allow the four basic operations of addition, subtraction, multiplication, and division to be carried out and the normal laws of algebra will apply. The real numbers $\mathbb{R}$ form a subset of $\mathbb{C}$, as the representation of a real number a as a complex number is $a + 0i$ (Figure 11).

When we add or subtract two complex numbers $z = (a, b)$, $w = (c, d)$, we simply add or subtract the first and second entries as

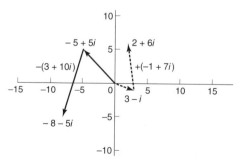

11. The Argand plane with addition and subtraction of complex numbers.

the case may be to give $(a, b) \pm (c, d) = (a \pm c, b \pm d)$. If we make use of the symbol i, we have, for example,

$$(3 - i) + (-1 + 7i) = (3 - 1) + i(-1 + 7) = 2 + 6i$$

and

$$(-5 + 5i) - (3 + 10i) = (-5 - 3) + (5 - 10)i = -8 - 5i$$

Addition of complex numbers corresponds to what is known as *vector addition* in the plane, where directed line segments (*vectors*) are added together, top to tail. We begin at the *origin*, which has coordinates $(0, 0)$, and in this example we lay down our arrow from the origin and place its tip at $(3, -1)$. To add the arrow represented by $(-1, 7)$, we begin at $(3, -1)$ and travel 1 unit to the left in the horizontal direction, which defines the *real axis*, and 7 units in the vertical direction, which is known as the *imaginary axis*. The outcome of the addition is that we end at the point $(2, 6)$. Similarly, subtraction of complex numbers corresponds to subtraction of vectors in the plane.

Before continuing, bear in mind that the adjective 'imaginary' is just a word and we should not read too much into it. After all, a complex number $a + bi$ can be thought of simply as a pair of real numbers (a, b), in much the same way as a fraction a/b can be represented by an integer pair (a, b). The validity of the complex numbers then comes down to whether or not the system is free from contradiction, which it is, while the importance of $\mathbb{C}$ stems from its effectiveness as an arena for calculation. Leaving the monorail of the real line and passing into the Argand plane allowed mathematics and physics to blossom in a way that would otherwise have been impossible.

Multiplication of two complex numbers z and w now comes quite easily. We simply expand the product of z and w using the distributive law, replace each instance of i^2 by -1, and then collect the real and imaginary parts together:

$$zw = (a + bi)(c + di) = ac + adi + bic + bidi$$

$$= (ac + bdi^2) + adi + bci;$$

$$\therefore zw = (ac - bd) + (ad + bc)i. \qquad (20)$$

But how can we divide one complex number by another? The hint comes through the notion of conjugation, which arises in relation to the quadratic formula.

The form of the solutions of a quadratic equation with rational coefficients, as provided by the quadratic formula, comes as a pair of numbers of the form $r = p + \sqrt{q}$ and $\bar{r} = p - \sqrt{q}$, where p and q are rational. We call r and $\bar{r}$ a *conjugate pair*. Conjugates can be used to simplify an expression with a square root in the denominator so that the denominator becomes simply an integer. This process, known as *rationalizing the denominator*, is brought about by multiplying the fraction top and bottom by the conjugate of the denominator. The cross-terms involving the square root then cancel out, leaving the simplified denominator. For example, we rationalize the denominator of $3/(5 - 2\sqrt{2})$ as follows:

$$\frac{3}{5 - 2\sqrt{2}} \cdot \frac{(5 + 2\sqrt{2})}{(5 + 2\sqrt{2})} = \frac{15 + 6\sqrt{2}}{5^2 + 10\sqrt{2} - 10\sqrt{2} - 4(\sqrt{2})^2}$$

$$= \frac{15 + 6\sqrt{2}}{25 - 8} = \frac{15}{17} + \frac{6}{17}\sqrt{2}.$$

The key property of the conjugate, $\bar{r}$, of r is that the product $r\bar{r}$ falls back into the field of the rationals and the irrational *surd* of $\sqrt{2}$ in the denominator disappears. The analogue for a complex number $z = a + bi$ would be another complex number $\bar{z}$ such that $z\bar{z}$ was purely real, which is to say the imaginary part of the product would be 0 and so the imaginary unit i in the denominator vanishes. A number with this qualification is the one that results from reflecting z in the real axis to give the *complex conjugate*, $\bar{z} = a - bi$:

$$z\bar{z} = (a + bi)(a - bi) = a^2 - abi + abi - b^2i^2 = a^2 + b^2. \qquad (21)$$

Note that $z\bar{z}$ is not only real, it is also non-negative and indeed will always be positive except if $a = b = 0$, which is to say $z = 0$. Note also that (21) tells us that $z\bar{z}$ is the square of the distance of $z = a + bi$ from the origin. In keeping with the absolute value notation that it generalizes, we write this distance, known as the *modulus* of z, as $|z|$ so that $z\bar{z} = |z|^2$.

We can now divide one complex number by another through use of the complex conjugate of the divisor—for example,

$$\frac{21 + i}{3 + 2i} = \frac{21 + i}{3 + 2i} \cdot \frac{3 - 2i}{3 - 2i} = \frac{63 - 42i + 3i - 2i^2}{3^2 + 2^2} = \frac{(63 + 2) - 39i}{9 + 4}$$
$$= \frac{65 - 39i}{13} = 5 - 3i.$$

A quadratic equation may have either 0, 1, or 2 solutions, according as the discriminant $\Delta = b^2 - 4ac$ is less than, equal to, or greater than 0. The quadratic formula provides these solutions and imaginary numbers do not need to be called upon to find them. However, once x^3 appears on the stage, complex numbers arise naturally in the course of the associated calculations, even when the solutions are all real numbers. The field of complex numbers was one of the biggest surprises ever to emerge from mathematics, second only to the discovery of irrational numbers by the Pythagoreans some two thousand years earlier.

Complex numbers allow us to find the roots of a quadratic in the case where the discriminant $\Delta < 0$. For example, let us solve $x^2 - 10x + 41 = 0$. In this case

$$\Delta = b^2 - 4ac = (-10)^2 - 4(1)(41) = 100 - 164 = -64 = 64i^2;$$

hence $\pm\sqrt{\Delta} = \pm 8i$ and so we obtain

$$x = \frac{-b \pm \sqrt{\Delta}}{2a} = \frac{10 \pm 8i}{2} = 5 + 4i \text{ or } 5 - 4i. \qquad (22)$$

There is a general point to take away from (22). The complex roots of a quadratic with real coefficients are always going to form a conjugate pair, $c \pm di$. As we shall see in the upcoming section, this principle applies equally to polynomials of higher degree.

The complex conjugate has the nice property of commuting with arithmetic operations: the conjugate of a sum, difference, product, or quotient is the sum, difference, product, or quotient as the case may be of the conjugates: in symbols, $\overline{z \pm w} = \bar{z} \pm \bar{w}$, $\overline{zw} = \bar{z}\,\bar{w}$, and $\overline{(z/w)} = \bar{z}/\bar{w}$. Each of these identities can be verified by calculating both sides of the corresponding equation for two arbitrary complex numbers $z = a + bi$ and $w = c + di$ and noting that the outcome is the same for both. For instance, using our multiplication rule (20), we see that

$$\overline{zw} = (ac - bd) - (ad + bc)i = (a - bi)(c - di) = \bar{z}\,\bar{w}.$$

For division, first note that for any non-zero complex number w we have $\overline{(w/w)} = \bar{1} = 1$. Now, since division by w means multiplication by its reciprocal, we may apply the result for the conjugate of a product to obtain

$$1 = \bar{1} = \overline{w \cdot \frac{1}{w}} = \bar{w} \cdot \overline{\left(\frac{1}{w}\right)} \quad \text{so that} \quad \overline{\left(\frac{1}{w}\right)} = \frac{1}{\bar{w}};$$

in general, then,

$$\overline{\left(\frac{z}{w}\right)} = \overline{\left(z \cdot \frac{1}{w}\right)} = \bar{z} \cdot \overline{\left(\frac{1}{w}\right)} = \bar{z} \cdot \frac{1}{\bar{w}} = \frac{\bar{z}}{\bar{w}}.$$

We shall exploit these relationships in the next section to show that the roots of polynomials with real coefficients come in conjugate pairs. However, another consequence is that

$$|zw| = |z| \cdot |w|, \tag{23}$$

and similarly the modulus of a quotient is the quotient of the moduli. To see (23) is now easy—since the quantities involved are non-negative real numbers, we need only check that their squares

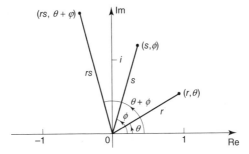

12. Multiplication of complex numbers in polar form.

are equal, which is the case:

$$|zw|^2 = (zw)(\overline{zw}) = zw\overline{z}\,\overline{w} = z\overline{z}\,w\overline{w} = |z|^2 \cdot |w|^2.$$

The rule (23) leads to another representation of complex numbers, in the form $z = (r, \theta)$, where r is the modulus of z; this places z somewhere on the circle centred at the origin O with radius r, and θ is the *angle* between the real axis and the ray Oz. This is called the *polar form* of z and is useful when dealing with powers and roots, for, in polar form, multiplication of $z = (r, \theta)$ and $w = (s, \phi)$ (Figure 12) follows the rule

$$zw = (r, \theta)(s, \phi) = (rs, \theta + \phi). \tag{24}$$

In particular, it follows through n-fold application of (24) that $z^n = (r^n, n\theta)$. That the moduli r and s multiply when we take a product follows from (23), and that the polar angles *add* can be shown by writing z in Cartesian form using trigonometric functions and applying the so-called double-angle formulae for cosine and sine.

Factorization of polynomials

The Factor Theorem tells us that if r is a root of a polynomial $p(x)$ of degree n then we may factorize $p(x)$ as $p(x) = (x - r)q(x)$, where $q(x)$ has degree $n - 1$. Repetition of this process leads, therefore, to the conclusion that the number of solutions to the equation $p(x) = 0$ is at most n. Indeed, allowing for repeated and complex roots, there will always be exactly n roots of an nth-degree polynomial. For example, the equation $x^2 + 1 = 0$ clearly has no real solutions but it does have two imaginary ones, they being $\pm i$, while the one and only solution of $x^2 - 2x + 1 = 0$ is $x = 1$. This apparent lack of a second root, however, is down to the fact that the root 1 is repeated, in that the factorization of $x^2 - 2x + 1$ is $(x-1)(x-1)$. We shall explore a little further.

The Factor Theorem sometimes gives us a foot in the door when it comes to solving equations featuring cubic and higher powers of the unknown x. If, for example, we can find one root a of a cubic polynomial $p(x)$, then the equation $p(x) = 0$ can be written as $(x - a)q(x) = 0$, where $q(x)$ is quadratic. The remaining solutions will then be those of the quadratic equation $q(x) = 0$ and so all solutions can be found. Indeed, it is possible to find all the rational roots of *any* polynomial equation $p(x) = 0$ with rational coefficients, for we may proceed as follows.

First, we may 'clear the denominators': by multiplying the polynomial by the product A of all the denominators, we arrive, through cancellation within each coefficient, at another polynomial with integer coefficients. This new polynomial has the same roots as the original because, since $A \neq 0$, for any value of x we have $Ap(x) = 0$ if and only if $p(x) = 0$. Therefore there is no loss of generality in assuming that $p(x)$ has the form

$$p(x) = a_0 + a_1 x + \ldots + a_n x^n,$$

where $a_0, a_1, \ldots, a_n$ lie in $\mathbb{Z}$, $n \geq 1$, and $a_n \neq 0$.

We are hunting for all the rational roots, if any, of $p(x)$ so let us suppose that $p(a/b) = 0$, where a and b are integers, with $b \neq 0$ of course. What is more, we may assume that the fraction a/b has been cancelled to lowest terms, so that 1 is the greatest common divisor of a and b. We will take for granted here a slight generalization of Euclid's Lemma, that being that if r and s have no common factor but $r|st$, then $r|t$.

Substituting $x = a/b$ in $p(x)$ and multiplying through by b^n to clear denominators we get the following, upon cancelling the powers of b that arise:

$$a_0 b^n + a_1 a b^{n-1} + a_2 a^2 b^{n-2} + a_3 a^3 b^{n-3} + \ldots + a_n a^n = 0. \qquad (25)$$

Now, if we divide the LHS by a, (25) tells us that the outcome is $0/a = 0$. However, a is an explicit factor of every term after the first: for example, $a_1 a b^{n-1}/a = a_1 b^{n-1}$ is an *integer*. It follows that the first term, $a_0 b^n$, must also be a multiple of a, otherwise the LHS could not equal the integer 0 when divided by a. However since a and b are relatively prime, the same is true of a and b^n as b and b^n have the same prime factors. We conclude, therefore, by the generalization of Euclid's Lemma, that *a must be a factor of the constant term* a_0 of $p(x)$. This means that the numerator a can be found among the factors (positive and negative) of a_0.

Similarly, we can narrow the search for the denominator b of our rational root. Since b appears explicitly as a factor of each term on the LHS of (25) apart from the final one, $a_n a^n$, it follows that b is a factor of *every* term on the LHS, including $a_n a^n$. But once again, a and b have no common factor, so the same is true for a^n and b. Therefore *b must be a factor of the leading coefficient* a_n of $p(x)$.

In conclusion, we have the *Rational Root Theorem*: for any rational root a/b of $p(x)$, the numerator $a|a_0$ and the denominator $b|a_n$.

We need to be aware of what this theorem does and does not say. It does *not* say that $p(x)$ has any rational roots, and it is possible that it may not. It does say, however, that *if* $p(x)$ has any rational roots, then those roots may be found within a *finite* set of rational numbers that we can list explicitly. It is then just a matter of testing each of these candidates in turn in order to find the complete list of rational roots of $p(x)$. It may be that none of these rational numbers a/b satisfy $p(a/b) = 0$. If that is the case, we still do not know any of the roots of $p(x)$, but we do know that they are not rational. And this is perfectly possible—for instance $x^2 - 2$ has $\pm\sqrt{2}$ as its roots, which are both irrational numbers.

As an example, let us find all the roots of $p(x) = 2x^3 + x^2 - 5x - 3$. For any rational root a/b of $p(x)$, we have by the Rational Root Theorem that a is a factor of the constant term, -3, and b is a factor of the leading coefficient, 2, so that a/b is equal to or is the negative of some member of the set $\{1, 3, \frac{1}{2}, \frac{3}{2}\}$. This gives us eight possibilities to test: for example, $p(1) = 2 + 1 - 5 - 3 = -5 \neq 0$. One candidate does, however, pass the test:

$$p\left(-\frac{3}{2}\right) = 2\left(-\frac{3}{2}\right)^3 + \left(-\frac{3}{2}\right)^2 - 5\left(-\frac{3}{2}\right) - 3$$

$$= -\frac{27}{4} + \frac{9}{4} + \frac{30}{4} - \frac{12}{4} = \frac{39 - 39}{4} = 0.$$

By the Factor Theorem, $(x - (-\frac{3}{2})) = x + \frac{3}{2}$ is a factor of $p(x)$ and, by the same token, so is any non-zero multiple of this factor: multiplying our factor by 2 allows us to express $p(x)$ as

$$p(x) = 2x^3 + x^2 - 5x - 3 = (2x + 3)(ax^2 + bx + c)$$

$$= 2ax^3 + (3a + 2b)x^2 + (3b + 2c)x + 3c.$$

Equating coefficients immediately gives us $a = 1$, $c = -1$, and $3 + 2b = 1$ so that $b = -1$. (Equally, $3b - 2 = -5$ gives $b = -1$.) Hence the quadratic factor is $x^2 - x - 1$, the roots of which we may now find—indeed, we already have done so, for this is the quadratic that arose when we found the value of the Golden Ratio, which has irrational roots $(1 \pm \sqrt{5})/2$.

Both in theory and in practice, the Rational Root Theorem allows us to find all the rational roots of any polynomial with rational coefficients. As anticipated in Chapter 1, we can use it to give the important theoretical result that so vexed the mathematicians of ancient Greece, that being that if a positive integer k is *not* the nth power of another positive integer, then $\sqrt[n]{k}$ is irrational.

To this end, consider the polynomial $p(x) = x^n - k$. Putting $p(x) = 0$, we see that its roots are $\pm\sqrt[n]{k}$. If the positive root were rational, it would have the form a/b, where we may take $a, b > 0$. However, by the Rational Root Theorem we have that the numerator a is a factor of k, while b is a factor of 1, which is to say $b = 1$, and so $a/b = a$ and the root a is then a whole number. This shows that $\sqrt[n]{k}$ either is a positive integer or is irrational, but $\sqrt[n]{k}$ cannot be a non-integral fraction. In particular, numbers such as $\sqrt{2}$, $\sqrt{60}$, and $\sqrt[3]{25}$ are irrational.

The same can be said for numbers like our Golden Ratio, ϕ, for if this were not the case we would have $\phi = a/b$ and, using the expression for ϕ in (16), we could argue as follows:

$$\phi = \frac{1+\sqrt{5}}{2} = \frac{a}{b} \quad \text{and so} \quad \sqrt{5} = \frac{2a}{b} - 1 = \frac{2a-b}{b},$$

thereby giving the contradiction that $\sqrt{5}$ could also be expressed as a fraction, which we now know it cannot. Therefore ϕ is indeed irrational.

It should be noted, however, that the irrational numbers are *not* closed under either addition or multiplication. For example, $\sqrt{2}$, $2 - \sqrt{2}$, and $\sqrt{2} \pm 1$ are all irrational, yet

$$(2 - \sqrt{2}) + \sqrt{2} = 2, \quad (\sqrt{2} - 1)(\sqrt{2} + 1) = 2 - \sqrt{2} + \sqrt{2} - 1 = 1.$$

The *Fundamental Theorem of Algebra* asserts that any non-constant polynomial $p(x)$ has a root, λ, which may be real or complex. The theorem is true quite generally—the coefficients of $p(x)$ may be real or complex numbers. The theorem has a long history and is difficult to prove rigorously; no proof is completely algebraic, but rather there is always some element of a spatial, or, as it is called in mathematics, *topological*, argument involved.

Although not a purely algebraic result, the Fundamental Theorem of Algebra has important consequences for the nature and associated factorizations of polynomials, as we now explain.

Let $p(x)$ be a typical nth-degree polynomial, and suppose z is a complex root of $p(x)$ so that $a_0 + a_1 z + \ldots + a_n z^n = 0$. Take the conjugate of both sides of this equation. Of course, $\overline{0} = 0$ and, applying the rules that the conjugate of a sum is the sum of the conjugates and the conjugate of a product is the product of the conjugates, we obtain

$$\overline{a_0} + \overline{a_1}\,\overline{z} + \ldots + \overline{a_n}\,\overline{z}^n = 0,$$

so that $\overline{z}$ is a root of this conjugated equation. If, however, the coefficients of $p(x)$ are real, then each $\overline{a_i} = a_i$, giving us the *Conjugate Root Theorem*: if z is a root of a polynomial $p(x)$ with real coefficients, then so is its conjugate $\overline{z}$.

The Fundamental Theorem of Algebra and the Factor Theorem now work in tandem to allow the complete factorization of a polynomial $p(x)$: taking the first root, r_1, allows us to factorize $p(x)$ as $(x - r_1)q(x)$, where $q(x)$ is a polynomial of degree $n - 1$. We can repeat the process n times in all until the quotient is simply a linear polynomial with root r_n, which will have the form $A(x - r_n)$ for some non-zero constant A. We then have a complete factorization of $p(x)$ into linear factors:

$$p(x) = A(x - r_1)(x - r_2)\ldots(x - r_n). \tag{26}$$

This is a general result: the coefficients of $p(x)$ may be complex, but this factorization still exists. The question remains as to how to find these n roots, but they certainly exist and there are no more. We do note, however, that some of the roots may equal one another. For example, suppose $p(x) = x^4(x - 1)(x - 2)^2$. The degree of $p(x)$ is 7 but $p(x)$ has only three distinct roots, which are 0, 1, and 2. The root 0 of $p(x)$ occurs in four factors and the root 2 in two factors, and so we say that 0 is a root of *multiplicity* 4 and, similarly, 2 is a root of multiplicity 2 for this polynomial.

There is one final interesting observation in this story in the case where the coefficients of $p(x)$ are all real numbers. By the Conjugate Root Theorem, the roots of $p(x)$ come in conjugate pairs, z and $\bar{z}$, and if z is not a real root, we get the second root, $\bar{z}$, 'for free'. This leads to the pair of factors in our factorization $(x-z)(x-\bar{z}) = x^2 - (z+\bar{z})x + z\bar{z}$. As we have already noted, if $z = a + bi$ then $z\bar{z} = a^2 + b^2$, which is a non-negative real number. Moreover, $z + \bar{z} = (a+bi) + (a-bi) = 2a$ is equally real. That is,

$$(x-z)(x-\bar{z}) = x^2 - 2ax + (a^2 + b^2).$$

We conclude that any polynomial with real coefficients can be factorized into a sequence of linear and quadratic factors *with real coefficients*. The quadratic factors $q(x)$ are *irreducible*, meaning that $q(x)$ cannot be factorized into linear factors over the reals (but, of course, if we allow complex numbers, we can split $q(x)$ into a pair of linear factors using its conjugate root pair). This is the way in which the Fundamental Theorem of Algebra is often expressed and was the form given by the renowned German mathematician Carl Friedrich Gauss (1777–1855) in his PhD thesis of 1799.

Solution of the cubic

We have made much progress, so let's take stock of where we stand on the problem of solving the cubic equation with rational (or equivalently integer) coefficients, $p(x) = ax^3 + bx^2 + cx + d = 0$. From (26) we know that $p(x) = 0$ has three (not necessarily distinct) roots. By the Conjugate Root Theorem, these three roots either are all real numbers or consist of two complex conjugate roots and one real root. In any event, there is at least one real root to find and if we can identifiy one root, r, we can by the Factor Theorem reduce the problem to solving an equation of the form

$(x - r)q(x) = 0$, where $q(x)$ is a quadratic. We could then find the roots of $q(x)$ and so completely solve our cubic.

What is more, we may use the Rational Root Theorem to find all of the rational roots of our equation. However, if it transpires that there are no rational roots, we still have no technique for finding a real root of our cubic.

We shall mimic our successful approach for solving the quadratic. Once again, we may assume that the leading coefficient $a \neq 0$, and divide through our equation by a to produce an equivalent monic polynomial equation. Hence it is enough to consider equations of the form $x^3 + ax^2 + bx + c = 0$. (The symbols a, b, and c here stand once again for general integer coefficients and are not the same as those of the original equation.) Taking our lead from the quadratic case, the next step is to substitute $x = y + t$. By making a cunning choice for t, we should be able to produce an equivalent equation without any term in y^2. Applying the binomial expansion for the power $n = 3$, we obtain

$$(y + t)^3 + a(y + t)^2 + \ldots = 0$$

and so

$$y^3 + (3t + a)y^2 + \text{(terms in } y \text{ and constants)} = 0.$$

Therefore, if we put $t = -a/3$ the outcome will be a monic cubic in y with no term in y^2. It follows that we will be able to solve any cubic provided that we learn how to solve a *depressed cubic*, which is to say one of the form $x^3 + dx + e = 0$, for the general cubic can be reduced to one of this kind.

The quickest passage to the solution is now through use of Vieta's substitution, $x = v + s/v$, where s is chosen to simplify the resulting equation into a quadratic. Making the substitution as suggested and applying the Binomial Theorem with $n = 3$ gives the following as the LHS:

$$\left(v + \frac{s}{v}\right)^3 + d\left(v + \frac{s}{v}\right) + e = v^3 + 3sv + \frac{3s^2}{v} + \frac{s^3}{v^3} + dv + \frac{ds}{v} + e$$

$$= v^3 + (3s + d)v + (3s + d)\frac{s}{v} + \frac{s^3}{v^3} + e.$$

We can now kill two algebraic birds with one stone by putting $s = -d/3$, for then *both* the terms in v and in $1/v$ vanish, leaving the equation

$$v^3 - \frac{d^3}{27v^3} + e = 0, \quad \text{whence } v^6 + ev^3 - \frac{d^3}{27} = 0.$$

Finally, we have a quadratic in v^3, meaning that by substituting $z = v^3$ we reduce the problem to the quadratic equation $z^2 + ez - (d/3)^3 = 0$. We now solve this equation to find z, take cube roots to get v, next recover y from $y = v - d/3v$, and finally we have our solution $x = y - a/3$.

Let's apply this process to an example:

$$p(x) = x^3 - 3x^2 + 6x + 8 = 0.$$

To get a depressed cubic we put $x = y + t$, where $t = -a/3 = -(-3)/3 = 1$, so we put $x = y + 1$, which gives

$$(y + 1)^3 - 3(y + 1)^2 + 6(y + 1) + 8$$
$$= (y^3 + 3y^2 + 3y + 1) - (3y^2 + 6y + 3) + (6y + 6) + 8 = 0$$
$$\Rightarrow y^3 + 3y + 12 = 0.$$

Next we use the Vieta substitution, $y = v + s/v = v - d/3v = v - 3/3v = v - 1/v$, which gives

$$\left(v - \frac{1}{v}\right)^3 + 3\left(v - \frac{1}{v}\right) + 12 = v^3 - 3v + \frac{3}{v} - \frac{1}{v^3} + 3v - \frac{3}{v} + 12 = 0$$

$$\Rightarrow v^3 - \frac{1}{v^3} + 12 = 0 \Rightarrow v^6 + 12v^3 - 1 = 0.$$

Putting $z = v^3$ gives the quadratic

$$z^2 + 12z - 1 = 0$$

$$\Rightarrow z = \frac{-12 \pm \sqrt{144+4}}{2} = \frac{-12 \pm \sqrt{148}}{2}$$

$$= \frac{-12 \pm \sqrt{4 \times 37}}{2} = \frac{-12 \pm 2\sqrt{37}}{2} = -6 \pm \sqrt{37}.$$

Taking the root $\sqrt{37} - 6$, we get that $v = \sqrt[3]{\sqrt{37} - 6}$ (the alternative choice leads to the same set of roots). Continuing, we may now find a real root:

$$x = y + 1 = v - \frac{1}{v} + 1 = \frac{v^2 + v - 1}{v}.$$

So, finally, we have our real root:

$$x = \frac{(\sqrt{37}-6)^{2/3} + (\sqrt{37}-6)^{1/3} - 1}{(\sqrt{37}-6)^{1/3}} \approx -0.8589.$$

Returning to the original polynomial $p(x) = x^3 - 3x^2 + 6x + 8$, we may check that $p(-1) = -2 < 0 < 8 = p(0)$. Since the graph of the polynomial is a continuous curve, $p(x)$ must therefore take on every possible value between -2 and 8 as x increases from -1 to 0. In particular, there must be a value of x in the interval $-1 < x < 0$ such that $p(x) = 0$. What is more, the values of $p(-1)$ and $p(0)$ suggest that this root will lie closer to -1 than to 0. The previous calculation confirms this and tells us precisely where that root lies.

Associated with a cubic, or indeed with any polynomial, there is a quantity Δ, called the *discriminant*, that determines the nature of its roots. In the case of a quadratic polynomial, $ax^2 + bx + c$, we have $\Delta = b^2 - 4ac$. Without giving the general definition here, suffice it to say that Δ is a certain product of squares of the differences of the roots. (For a monic quadratic with roots r and s, we can check that $\Delta = (r-s)^2$.) Since the discriminant has the product of the differences of the roots as a factor, it follows that $\Delta = 0$ if and only if two of the roots are identical, which is to say the polynomial has a repeated root. In the real coefficient cubic case, we can say more, in that $p(x)$ has three real roots if $\Delta \geq 0$ but if $\Delta < 0$ the polynomial has just one real root and a pair of complex conjugates for the other roots.

The definition of Δ is theoretically useful as it reveals facts such as those just mentioned, but since it is expressed in terms of the roots, which are unknown, it is not a useful form for calculation. However, the value of Δ can be expressed in terms of the coefficients of the polynomial, although the expression is very complicated for high powers. For the cubic $ax^3 + bx^2 + cx + d$, it transpires that

$$\Delta = b^2c^2 - 4ac^3 - 4b^3d - 27a^2d^2 + 18abcd.$$

In our example, $a = 1$, $b = -3$, $c = 6$, and $d = 8$ and so

$$\Delta = (-3)^2(6^2) - 4(1)(6^3) - 4(-3)^3(8) - 27(1^2)(8^2) + 18(1)(-3)(6)(8)$$

$$= -3996 < 0;$$

hence the real root that we found is the only one. However, as the pioneers of the Italian Renaissance discovered, in the case of a positive discriminant, the substitution method leads to expressions involving complex numbers even though all three roots are real.

We close with a few words on *quartic* and *quintic* equations, which are respectively polynomial equations of degree four and five. Methods for solving both cubic and quartic equations appeared together in the book *Ars Magna*, written by Gerolamo Cardano in 1545. The method for solving the quartic was devised by Cardano's student Lodovico Ferrari (1522–65). First, a simple substitution reduced the problem to that of the monic depressed quartic, which has the form $x^4 + ax^2 + bx + c = 0$. At this point Ferrari devised an ingenious substitution that in principle allowed solution of the problem. The difficulty that remains, however, is that the newly introduced variable y has to be chosen so that the discriminant of a certain associated quadratic is 0. In order for this to hold, y has to be the solution of a certain cubic equation. In this way, Ferrari had shown that provided that cubic equations could be solved, then so could fourth-degree equations.

Eventually, though, the line of manipulations and substitutions devised to find roots of polynomials becomes exhausted. The Ruffini–Abel Theorem proved that there is no formula for the roots of a fifth-degree (or higher) polynomial equation in terms of the coefficients of the polynomial, using only the usual algebraic operations (addition, subtraction, multiplication, and division) and application of radicals (square roots, cube roots, etc.). The modern development of the theory of roots of equations, however, stems from the efforts of Évariste Galois (1811–32), whose work led to an entirely new branch of abstract algebra known as group theory, which has dominated research in the subject ever since.

Chapter 6
Algebra and the arithmetic of remainders

This chapter features a new type of algebra, the arithmetic of remainders, which is both an ancient topic and one that has found a major contemporary application in Internet cryptography. We first say a little more about abstract algebra, as it forms the backdrop to the particular example types that you will meet in the remainder of the book.

Groups, rings, and fields

The integers under the operation of addition, $(\mathbb{Z}, +)$, are a key example of what is known as a *group*. A group consists of an underlying set, which in the case of the integers is $\mathbb{Z}$, coupled with an operation, addition (+) in this case, which allows two members of the set, a and b say, to interact and produce another member, $a + b$, of the same set. A group operation must be a *binary operation*, like addition, meaning that it involves *two* elements of the set. What is more, for a binary operation to be a group operation we insist further that the operation satisfies three particular conditions, all of which hold for integer addition: the operation, +, must be associative, i.e. $a + (b + c) = (a + b) + c$ for any three members a, b, c of the set; there must be an *identity element*, denoted by 0, which has the property that $a + 0 = 0 + a = a$ always holds; and, finally, each member a of the set must have an *inverse* element, denoted here by $-a$, that reverses the effect of adding a in the sense that $a + (-a) = (-a) + a = 0$, the identity element.

Integer addition also satisfies the commutative law in that $a + b = b + a$. Commutativity is not, however, part of the general definition of a group, but when the operation of a group G does respect the commutative law, we say that G is an *abelian group*, a term derived from the surname of Niels Abel (1802–29), who gave his name to the Ruffini-Abel Theorem mentioned in Chapter 5 in relation to the insolvability of fifth-degree polynomial equations.

From Chapter 7 onwards we will meet other examples of sets, in particular sets of matrices, which form groups with operations that are completely different from ordinary addition. However, because these new operations still satisfy the group axioms, any general results about groups will hold in these contexts as well, and this gives a pointer as to why groups and other abstract algebras are studied in full generality.

The payoff from this approach stems from the fact that whatever theorems and relationships come to light in the context of abstract algebra then apply to any particular algebraic object that obeys the rules in question. Mathematicians often seek out the widest setting in which key theorems hold, as that not only increases their scope but also gives a clearer understanding as to why they are true. For these reasons, you will find that textbooks on abstract algebra often refer to *semigroups* and *groups*, which are algebras with a single associative operation, while *rings* and *fields*, which are notions that we are about to introduce, are algebras with two operations linked via the distributive law. And there are more: *lattices* are algebras with an ordered structure, while *vector spaces* and *modules* are algebras where the members can be multiplied by scalar quantities from other fields or rings. Algebras themselves are also studied collectively. This holistic approach is known as *universal algebra* and its practitioners search for theorems that are valid for algebras of all kinds.

The integers under addition are an abelian group that we have been dealing with since childhood, which is why the group properties in this context are natural to all of us. However, right from the beginning, we work with *two* operations, addition and multiplication, when dealing with $\mathbb{Z}$. For this reason we need to consider algebras like the integers and like the rational numbers, $\mathbb{Q}$, which have a pair of binary operations, linked by the distributive law.

A *commutative ring* is a type of algebra, such as the integers $\mathbb{Z}$, which possesses two operations, denoted by $+$ and $\cdot$. Under the operation denoted by $+$, the set forms an abelian group. The multiplication operation too is associative and commutative but a ring is also required to satisfy the distributive law of addition, $+$, over multiplication, $\cdot$, namely that $a \cdot (b + c) = a \cdot b + a \cdot c$. A different kind of example of a commutative ring is the ring $\mathbb{P}$ of all polynomials with, let us say, real coefficients, under the operations of polynomial addition and multiplication.

A *field* is a special type of commutative ring, an example being the rational numbers $\mathbb{Q}$, where there is also a multiplicative identity, often denoted by 1 (an attribute that $\mathbb{Z}$ also possesses; for this reason $\mathbb{Z}$ is known as a *unital ring*), and every non-zero member a of a field has a *multiplicative inverse*, denoted by either a^{-1} or $1/a$, such that $a^{-1} \cdot a = 1$.

Our two extended number systems of the real numbers, $\mathbb{R}$, and the complex numbers, $\mathbb{C}$, both pass the test that earns them the title of a field. On the other hand, the collection $\mathbb{P}$ of all polynomials with real coefficients form a commutative unital ring that is not a field.

In Chapter 2 we proved results such as $a \times 0 = 0$ and that the additive inverse, $-a$, of a is unique. Although we were thinking of the letter a as an integer when conducting those arguments, the symbol a could have stood for any member of an arbitrary commutative ring, as the argument rests only on laws that apply in that context. It follows that these results apply in particular to the ring of polynomials, $\mathbb{P}$. Moreover, the Binomial Theorem holds in any commutative ring, so the symbols x and y in the binomial $(x + y)^n$ can stand for polynomials and the conclusion of the theorem is equally valid.

Although these are quite simple results, $\mathbb{P}$ has more in common with $\mathbb{Z}$, in that a form of the Euclidean Algorithm works for polynomials as well: for any polynomials $p(x)$ and $q(x)$, we may find a unique monic polynomial, $g(x)$, which is a gcd of $p(x)$ and $q(x)$ in the sense that any other such common factor of $p(x)$ and $q(x)$ is itself a factor of $g(x)$. This guarantees that the rings $\mathbb{Z}$ and $\mathbb{P}$ share other more subtle algebraic properties concerning their internal make-up, known as their *ideal* structure, which will not be elaborated on further here but is central to their study.

We shall shortly be leaving these abstract considerations for the time being, but the idea to take forward is that various types of algebras have been defined where the collections under consideration need to satisfy a special set of algebraic rules. The reason why one particular arrangement of rules is the focus of attention is because certain important yet different collections of algebraic objects have been found to share these properties, and for that reason it is worthwhile studying particular types of algebra in full generality, groups, rings, and fields being three cases in point.

Before moving on, however, we make one more observation that particularly applies to the integers. Early in the piece we proved that any product involving zero is equal to zero: $a \times 0 = 0$. Is the converse true? In other words, *if* a product of integers $ab = 0$, may we infer that at least one of the factors, a or b, must be 0?

In the case of the ring of integers, the answer is yes, and a commutative unital ring with this property is called an *integral domain* in recognition of this integer-like property. We can see this is true by observing that $\mathbb{Z}$ may be regarded as being embedded in its *field of fractions*, which is $\mathbb{Q}$. If we have $ab = 0$ in $\mathbb{Z}$ then $ab = 0$ in $\mathbb{Q}$ as well, and then either $a = 0$ or, if not, we may multiply both sides of this equation by a^{-1} to obtain

$$(a^{-1}a)b = a^{-1} \cdot 0 \Rightarrow 1 \cdot b = 0 \quad \text{and so } b = 0;$$

hence if $ab = 0$ then at least one of a and b is 0.

Although we cannot generally divide within an integral domain such as the integers, we can freely cancel because the characteristic property of an integral domain is just what is needed to guarantee this. Suppose we have $ab = ac$ in an integral domain with $a \neq 0$. Then $ab - ac = a(b - c) = 0$, and since a is not 0, it follows that $b - c$ is zero, which is to say $b = c$, and so we may cancel the common factor a from both sides of $ab = ac$. Not every commutative ring enjoys this feature. As we shall see in the next section, there are commutative rings that are not integral domains (and so cannot be embedded into a field). Indeed, the members of these rings can be integers—the addition and multiplication operations, however, are a little different.

Modular arithmetic: rules of engagement

We now set out to establish the algebraic rules for the *ring of integers modulo n*. Modular arithmetic is often called clock arithmetic, as we imagine a clock face with n numbers, $0, 1, \ldots, n - 1$, and we do our arithmetic on that, meaning that whenever you go past $n - 1$ you return again to 0 in a fashion reminiscent of a clock, or of a turnstile, or of cycles of a calendar. That all sounds quite simple and inconsequential, but the opposite is true. Modern cryptography is based on clock arithmetic and if it were all simple, the codes that rely on it would be simple to crack, but they are anything but. The erratic and unpredictable way that remainders arise under division lies at the heart of it all. We begin by identifying where two integers are essentially the same from the point of view of our arithmetic clock.

We call two integers a and b equivalent or *congruent modulo n* if $n|a - b$, which is to say that $a - b = kn$ for some integer k or, if you prefer, $a = b + kn$. Intuitively, a and b are equivalent if they represent the same time on the n-hour clock. We denote equivalence modulo n by the equation $a \equiv b \pmod{n}$.

A third way of phrasing this notion, which is perhaps the most important, is that $a \equiv b \pmod{n}$ if and only if a and b *leave the same remainder when divided by n*. For example, $22 \equiv 40 \pmod{6}$ as both 22 and 40 leave the remainder 4 when divided by 6.

The congruence sign is meant to suggest something akin to equality, and this is borne out by the basic observations that $a \equiv a \pmod{n}$, that if $a \equiv b \pmod{n}$ then it is equally the case that $b \equiv a \pmod{n}$, and, as with ordinary equality, that the congruence sign has the *transitive property* in that if $a \equiv b \pmod{n}$ and $b \equiv c \pmod{n}$ then $a \equiv c \pmod{n}$, as all three numbers will necessarily leave the same remainder when divided by n. A consequence of these three properties is that congruence mod n partitions the integers into n *equivalence classes*, as they are called. For example, for $n = 4$ the four classes, are

$$\{\ldots, -8, -4, 0, 4, 8, 12, \ldots\}, \quad \{\ldots, -7, -3, 1, 5, 9, 13, \ldots\},$$

$$\{\ldots, -6, -2, 2, 6, 10, 14, \ldots\}, \quad \{\ldots, -5, -1, 3, 7, 11, 15, \ldots\},$$

so that $-3 \equiv 9 \pmod{4}$, $24 \equiv 0 \pmod{4}$, and so on. The n classes of equivalence modulo n are often written as $[0], [1], [2], \ldots, [n-2], [n-1]$ and the representatives of each of these classes, $0, 1, 2 \ldots, n-1$, are called the *least residues* modulo n, which is to say they are the n possible remainders when an integer is divided by n. Every integer lies in one and only one of these n classes.

Modular arithmetic is all about manipulating these least-residue classes and, given that they represent infinite collections, that may sound daunting. However, you are well used to the idea that a fraction such as $\frac{2}{3}$ is equal to, or more accurately is equivalent to, any of an infinite number of fractions of the form $2m/3m$. As long as we are confident in the rules that govern fractions, this causes no trouble. The saving grace is that there is a unique representative of a fraction, that being the fraction cancelled to its lowest terms, which we tend to work with where possible. At times, however, fractions that are not fully cancelled down arise during calculations. In much the same way, the least residue is the smallest non-negative number in its equivalence class modulo n, and we shall normally work with these representatives while being aware that sometimes it may be convenient to use other representatives instead.

We will, however, need to know the algebraic rules that govern our new arithmetic, which is what we now go about establishing.

The notation $a \equiv b \pmod{n}$ is highly suggestive of ordinary equality, tempting us to indulge in the same kinds of algebraic manipulations we have worked with since our schooldays. If this were not the case, the choice of notation would not be appropriate. We will find, however, that the laws of modular arithmetic, although similar, are not identical to those of ordinary addition and multiplication.

We do not record complete proofs of all the results listed in this section. However, the claims, which all follow from the definition of congruence and elementary properties of number division, can be found in any textbook on number theory.

We begin with a simple observation, $(a + c) - (b + c) = a - b$, from which it follows that $a \equiv b \pmod{n}$ if and only if $a + c \equiv b + c \pmod{n}$, so that we may freely add any integer to both sides of a *congruence*, as these equations are often called (and so also subtract any integer). This fact ensures that whenever $a \equiv b \pmod{n}$ is true, we may replace $a + c$ by $b + c$ in any congruence equation modulo n.

Other nice properties of congruences now follow, which can be justified by arguments like the previous one. For instance, if $a_1 \equiv a_2 \pmod{n}$ and $b_1 \equiv b_2 \pmod{n}$, then $a_1 + b_1 \equiv a_2 + b_2 \pmod{n}$. This in effect now gives us an addition operation on the congruence classes, for it says that when we add two numbers modulo n, the class of the answer does not depend on which numbers are used to represent those classes in the sum. For example

$$15 + 11 \equiv 3 - 1 \equiv 2 \pmod{12} \quad \text{as } 15 \equiv 3 \text{ and } 11 \equiv -1 \pmod{12}.$$

We call this addition of classes via their representatives *addition modulo n* and write $(\mathbb{Z}_n, +)$ to denote the set of least residues $\{0, 1, 2, \ldots, n-1\}$ under the operation, $+$, of addition modulo n. The additive identity of this number system is 0, and the operation $+$ is commutative and associative. We thus have a fully fledged abelian group, as each $a \in \mathbb{Z}_n$ has $n - a$ as its inverse because $a + (n - a) = n \equiv 0 \pmod{n}$.

For multiplication, it follows in a similar fashion to addition that $a \equiv b \pmod{n}$ implies that $ac \equiv bc \pmod{n}$, and from this we deduce that $a_1 \equiv a_2 \pmod{n}$ and $b_1 \equiv b_2 \pmod{n}$ imply that $a_1 b_1 \equiv a_2 b_2 \pmod{n}$. Hence, as with addition, we may replace any number in an expression involving multiplication by any other from its congruence class modulo n and the result is congruent modulo n to the original. Moreover, congruence classes may be multiplied together through their representatives, and the outcome is independent of which representatives we adopt. What is more, commutativity, associativity, and distributivity of multiplication all hold in consequence of these laws being valid for the integers. In conclusion, we now have a commutative unital ring with a multiplicative identity 1, that ring being denoted by $(\mathbb{Z}_n, +, \times)$.

In the previous section, it was pointed out that the ring of integers is also an integral domain, giving multiplication in $\mathbb{Z}$ the cancellation property. Is that property also inherited by its 'image'

$\mathbb{Z}_n$? The answer is in general 'no' because if n is a composite number, $n = ab$ say, then neither a nor b is congruent to 0 modulo n but $ab = n \equiv 0 \pmod{n}$. As a consequence, we cannot cancel freely: for example, $15 \times 6 \equiv 11 \times 6 \equiv 18 \pmod{24}$ but we cannot cancel the common factor of 6 to conclude that $15 \equiv 11 \pmod{24}$, for that is clearly false.

An important class of exceptions is that of the rings $\mathbb{Z}_p$, where p is a prime, as here Euclid's Lemma comes to the rescue: if $ab \equiv 0 \pmod{p}$ then $p|ab$, so that $p|a$ or $p|b$, which is to say $a \equiv 0 \pmod{p}$ or $b \equiv 0 \pmod{p}$, which says exactly that $\mathbb{Z}_p$ is an integral domain. Indeed, that makes $\mathbb{Z}_p$ a field, for any *finite* integral domain F is a field. (This is because the cancellation law tells us that for any non-zero $a \in F$, the list of products of the form ab cannot have repeats and so, by finiteness, must exhaust the whole of F: in particular, one product ab must equal the multiplicative identity 1, and therefore a does indeed have an inverse, namely this number b.) Hence, for any prime p, the integral domain $\mathbb{Z}_p$ is a finite field with p elements. It turns out that there are other finite fields (indeed, there is exactly one finite field for every prime power p^n), but there are no others. This topic will be revisited in our final chapter.

Returning to the subject of the current chapter, the rings $\mathbb{Z}_n$, there is nonetheless a form of cancellation that is valid across any congruence sign. Let d denote the gcd of c and n. Then $ac \equiv bc \pmod{n}$ implies that $a \equiv b \pmod{n/d}$. For example, from the equation $24a \equiv 60b \pmod{93}$, we may wish to cancel the common factor of $c = 12$ in the congruence, while $n = 93 = 3 \times 31$. Hence the gcd of c and n is $d = 3$, and we conclude that $2a \equiv 5b \pmod{31}$. That is to say, you can be sure that $2a - 5b$ is a multiple of 31, but it is not necessarily a multiple of the larger, original modulus 93.

We close this section by drawing attention to one big advantage of arithmetic modulo n over the ordinary non-modular type, which is that since there are but n different possibilities to consider, important facts can often be verified simply by testing all the n numbers involved. For example, we now show that the sum of three squares, $a^2 + b^2 + c^2$, never has the form $8k + 7$.

To see this, instead of using the least residues $0, 1, \ldots, 7$ to represent the eight congruence classes, let us use the alternative set $-3, -2, -1, 0, 1, 2, 3, 4$, as this collection is simpler to deal with when squaring. Any square a^2 equals modulo 8 one of 0, 1, 4; for example, if $a \equiv -3 \pmod{8}$ then $a^2 \equiv (-3)^2 = 9 \equiv 1 \pmod{8}$. It follows that the sum of three squares modulo 8 is equal to a sum of three of the numbers (with repeats allowed) from the list of possibilities of 0, 1, 4. We find that under these rules we can generate the numbers 0, 1, 2, 3, 4, 5, and 6 (for example, $6 = 1 + 1 + 4$), but not 7. Therefore we conclude that for *any* three squares, $a^2 + b^2 + c^2 \not\equiv 7 \pmod{8}$. A famous theorem in number theory is that *any* non-negative integer is the sum of *four* squares. The preceding calculation shows, however, that there are infinitely many positive integers, namely $7, 15, 23, 31, \ldots$, that are not the sum of three.

Solving linear congruences

Having described the general algebraic structure of the ring $\mathbb{Z}_n$, we will now show how to solve linear equations in this ring: equations of the form $ax + c \equiv d \pmod{n}$. Of course, there is no difficulty in subtracting c from both sides of this equation, and so the real question is how do we solve congruences of the form $ax \equiv b \pmod{n}$?

To witness the range of behaviour offered by even the simplest congruences, consider the near-identical trio

$$3x \equiv 2 \pmod{6}, \quad 5x \equiv 2 \pmod{6}, \quad 4x \equiv 2 \pmod{6}. \qquad (27)$$

By testing each of the six possible values, we find that the first congruence has no solution at all, the second has a unique answer, 4, and the third equation has two solutions, namely 2 and 5.

The following theorem gives the number of solutions to $ax \equiv b$ (mod n). Let d stand for the gcd of a and n. There are no solutions to the congruence equation if d is not a factor of b, but if it is then there are d solutions.

This is consistent with what we have just observed about the three equations in (27). The first has no solution, as the gcd of the pair 3 and 6 is 3, which is not a factor of 2. In the second congruence, the gcd of 5 and 6 is 1, and of course 1 is a factor of 2 and our second equation indeed has exactly one solution. For the final equation, the relevant gcd is $d = 2$, the gcd of 4 and 6, and since 2|2 is certainly true, we expect and obtain two solutions here, all in accord with the general description provided by the theorem.

To see that we must have $d|b$ in order to have a solution, let us suppose that $ax \equiv b$ (mod n), whence $ax - b = kn$, say, or, making b the subject, $b = ax - kn$. It follows that any common factor of a and n must divide b also, and in particular this is true of d, the gcd of a and n.

As was implicitly shown in Chapter 3, a linear equation $ax = b$ in a field has a unique solution, $x = a^{-1}b$, so let's begin with the case where n is a prime, for here we know from the previous section that $\mathbb{Z}_n$ is indeed a field.

The solution is formally given by $x = a^{-1}b$, but the question remains as to how to find a^{-1}. Let us take an example, $6x \equiv 14$ (mod 31). Since 31 is prime and we are working in a field, we may indeed cancel and simplify to $3x \equiv 7$ (mod 31). The simplest way to solve this is now to add multiples of the modulus to the RHS until we can cancel the remaining coefficient of 3, and so we consider the sequence $7, 7 + 31 = 38, 38 + 31 = 69, \ldots$. We now get the answer from $3x \equiv 69$ (mod 31) and so $x \equiv 23$ (mod 31), which is to say that 23 is the unique least-residue solution to our original congruence.

In the case of a composite modulus, however, there may be more than one answer, as we saw with (27). However, the congruence can be solved by adding multiples of the modulus to the RHS until the coefficient of x has been cancelled down to 1. (The argument that justifies this claim is based on working the Euclidean Algorithm in reverse.) The full set of solutions then consists of the fundamental solution, x say, to which we may add multiples of $n' = n/d$ until we have the full suite of d solutions, as is illustrated in the next two examples.

For $4x \equiv 2 \pmod 6$, we have $d = \gcd(a, n) = \gcd(4, 6) = 2$. We may cancel the 2 to get $2x \equiv 1 \pmod 3 \Leftrightarrow 2x \equiv 4 \pmod 3 \Leftrightarrow x \equiv 2 \pmod 3$. The full set of least-residue solutions to the original congruence then consists of the fundamental solution, in this case $x = 2$, to which we may add $d - 1 = 2 - 1 = 1$ further solution, that being

$$x + n' = x + \frac{n}{d} = 2 + \frac{6}{2} = 2 + 3 = 5.$$

A more challenging example is

$$30x \equiv 24 \pmod{57}.$$

Here, $a = 30$, $b = 24$, $n = 57$, $d = \gcd(30, 57) = 3$ and so

$$n' = \frac{n}{d} = \frac{57}{3} = 19.$$

Since $3|24$, we cancel accordingly and obtain $10x \equiv 8 \pmod{19}$. We now have a prime modulus with a unique solution and we may freely cancel again to obtain $5x \equiv 4 \pmod{19}$. We look through the sequence $4 + 19n$ to find a multiple of 5:

4, $4 + 19 = 23$, $23 + 19 = 42$, $42 + 19 = 61$, $61 + 19 = 80$.

Hence $5x \equiv 80 \pmod{19}$, whence $x \equiv 16 \pmod{19}$. Since $d = 3$, we have three solutions all told, and since $n' = 19$ these are 16, $16 + 19 = 35$, and $35 + 19 = 54$.

Chapter 7
Introduction to matrices

Matrices represent the central algebraic vehicle for advanced computation throughout mathematics as well as in the physical and social sciences. Their introduction in the second half of the 19th century made mathematicians aware that there were important algebraic systems apart from number fields, and their study stimulated the growth of abstract algebra in the early 20th century.

Matrices and their operations

The word 'matrix' has a number of meanings in science as one kind of aggregation or another. Its meaning in mathematics is more akin to its use in the film *The Matrix*, where the characters inhabited a virtual world that was coded as an enormous binary array on a gigantic computer. Indeed, a matrix is simply a rectangular array of numbers of some kind: for example,

$$A = \begin{bmatrix} 9 & 0 & 3 & 1 & 4 \\ 2 & 8 & 4 & 7 & 0 \\ 1 & 1 & 5 & 4 & 4 \end{bmatrix}, \quad B = \begin{bmatrix} -1 & 0 & 4 & 2 & 0 \\ -1 & 4 & 3 & 2 & 3 \\ -1 & 0 & 6 & -1 & 12 \end{bmatrix}.$$

These two matrices, A and B, both have the same size in that each has 3 rows and 5 columns, but a matrix may have any number of rows, m, and columns, n. The matrices A and B are therefore 3×5 matrices; an $n \times n$ matrix is naturally called a *square* matrix.

There are some natural and simple operations that can be performed on matrices. First there is *scalar multiplication*, where all entries in a matrix are multiplied by a fixed number: I am sure the reader can write down the matrices $2A$ and $-3B$ without difficulty. Two matrices of the same size may be added (or subtracted) by adding (or subtracting) the corresponding entries to form a new matrix of the same size as the original pair. Indeed, we can form *linear combinations* of the form $aA + bB$ for any multipliers a and b. For example,

$$2A - 3B = \begin{bmatrix} 21 & 0 & -6 & -4 & 8 \\ 7 & 4 & -1 & 8 & -9 \\ 5 & 2 & -8 & 11 & -28 \end{bmatrix};$$

for instance, the entry at position (3, 1), meaning the third row and first column, is given here by $2(1) - 3(-1) = 2 + 3 = 5$.

This is all very simple, and it is fair to ask why we would want to look at this idea at all. What problems are we hoping to solve by introducing these arrays?

What makes matrices important is the way they are multiplied, for this is a completely new operation that arises in a host of problems involving many variables, which is a feature of complex, real-world situations. Before we move on to that it is worth noting, however, that the definition above allows matrix addition to satisfy the commutative and associative laws, as for $m \times n$ matrices A, B, and C,

$$A + B = B + A, \quad A + (B + C) = (A + B) + C.$$

Both these facts are immediate consequences of the corresponding laws holding for ordinary numbers. Also, for scalars a, b, and c we have the rules

$$(ab)A = a(bA); \quad a(A + B) = aA + aB, \quad (a + b)A = aA + bA. \tag{28}$$

None of this is surprising, but it does give notice that the laws of algebra can be applicable to algebraic systems other than various number fields such as $\mathbb{Q}$ and $\mathbb{C}$.

Matrix ideas have a very long history. The ancient Chinese text *The Nine Chapters on the Mathematical Art* (compiled around 250 BC) features the first examples of the use of array methods to solve simultaneous equations, including the concept of determinants (the subject of Chapter 9). The idea first appeared in Europe in Cardano's 16th-century work *Ars Magna*, mentioned in Chapter 5 in connection with the solution of the cubic.

From about 1850, the British mathematician James Sylvester did much to establish matrix and determinant theory as a branch of mathematics in its own right, along with his friend Arthur Cayley, who first introduced the formal notion of a group. In the 20th century, matrix theory arose irresistibly in many branches of mathematical physics, particularly in quantum mechanics.

To motivate matrix multiplication, we write our example of a 2×2 set of linear equations from Chapter 3 in matrix form. Since the problem is determined by the coefficients of the unknowns together with the right-hand sides of the equations, we isolate the *coefficient matrix* and write the pair of equations as a single matrix equation in the following fashion:

$$\begin{matrix} -2x + 5y = 34 \\ 3x + 4y = -5 \end{matrix} \Leftrightarrow \begin{bmatrix} -2 & 5 \\ 3 & 4 \end{bmatrix} \begin{bmatrix} x \\ y \end{bmatrix} = \begin{bmatrix} 34 \\ -5 \end{bmatrix}.$$

In order that our matrix equation represents a matrix product, we require that

$$\begin{bmatrix} -2 & 5 \\ 3 & 4 \end{bmatrix} \begin{bmatrix} x \\ y \end{bmatrix} = \begin{bmatrix} -2x + 5y \\ 3x + 4y \end{bmatrix}.$$

Regarding the RHS as the product of the two matrices on the left, we see that the first entry, $-2x + 5y$, is equal to -2, the first entry of the first row of the first matrix, times the first entry of the column matrix $\begin{bmatrix} x \\ y \end{bmatrix}$, plus the second entry, 5, of the first row times the second entry of the column matrix. On the other hand, $3x + 4y$ is the result of adding the first entry of the *second* row of the 2 × 2 matrix times the first entry, x, of the column plus the second entry of the second row times the second entry of the column. This is how matrix products are formed in general, as a sum of products of rows of the first matrix times columns of the second. For that to work, we need the length of the rows of the first matrix to match the length of the columns of the second. Another way of putting this is to say that the number of columns of the first matrix must equal the number of rows of the second.

Having said that, we make this precise as follows. Let A be an $m \times n$ matrix and B an $n \times k$ matrix. The product $C = AB$ will be an $m \times k$ matrix, which is to say the product has the same number of rows as A, the matrix on the left, and the same number of columns as the second matrix, B, on the right. A matrix with just one row is called a *row vector* and, similarly, a one-column matrix is a *column vector*.

The next example is of a 2 × 3 matrix, A, multiplied by a 3 × 2 matrix, B, resulting in a 2 × 2 product matrix:

$$AB = \begin{bmatrix} 1 & 2 & -1 \\ -1 & 0 & 1 \end{bmatrix} \begin{bmatrix} 1 & -2 \\ 0 & 3 \\ -1 & 0 \end{bmatrix} = \begin{bmatrix} 2 & 4 \\ -2 & 2 \end{bmatrix}.$$

For example, the entry in the top right-hand corner of the product is 4, which comes from taking the so-called *scalar product* (or *dot product*) of the first row of A and the second column of B:
$1 \times (-2) + (2 \times 3) + (-1 \times 0) = -2 + 6 - 0 = 4$.

Let us now compute the reverse product:

$$BA = \begin{bmatrix} 1 & -2 \\ 0 & 3 \\ -1 & 0 \end{bmatrix} \begin{bmatrix} 1 & 2 & -1 \\ -1 & 0 & 1 \end{bmatrix} = \begin{bmatrix} 3 & 2 & -3 \\ -3 & 0 & 3 \\ -1 & -2 & 1 \end{bmatrix}. \qquad (29)$$

This example shows that matrix multiplication spectacularly fails the commutative law. The matrices AB and BA are certainly unequal, for they do not even have the same dimensions. And this bad behaviour is not just a feature of oblong matrices. If you write down two square $n \times n$ matrices A and B, then both AB and BA will also be $n \times n$ matrices but they will otherwise probably be quite different from one another.

The general description of the matrix product is as follows. Let $A = (a_{ij})$ be an $m \times n$ matrix, meaning that the entry in row i and column j is a_{ij}. Similarly, let $B = (b_{ij})$ denote an $n \times k$ matrix. Then the product $C = AB$ has entries (c_{ij}), where c_{ij} is the *scalar product* of row i of A and column j of B, which is the number

$$c_{ij} = a_{i1}b_{1j} + a_{i2}b_{2j} + \cdots + a_{in}b_{nj}. \qquad (30)$$

The equation (30) says that c_{ij} equals the first entry of row i of A times the first entry of column j of B, plus the second entry of row i times the second entry of column j, and so on.

There is something slightly peculiar about the product BA in (29). If we label the row vectors of BA from top to bottom as $\mathbf{r}_1$, $\mathbf{r}_2$, and $\mathbf{r}_3$, you will notice that $\mathbf{r}_1 + \frac{2}{3}\mathbf{r}_2 + \mathbf{r}_3 = \mathbf{0}$, where $\mathbf{0} = (0, 0, 0)$ is the *zero vector*. We say that the row set of BA is *dependent*, in that one row can be expressed in terms of the others: for example, here we might write $\mathbf{r}_1 = -\frac{2}{3}\mathbf{r}_2 - \mathbf{r}_3$. This is an instance of a general phenomenon, as we next explain.

For any matrix M, the *row rank* of M is the maximum number of rows we can locate in the matrix that form an *independent set*, meaning that no row is equal to a *linear combination*, which means a sum of multiples, of the other rows. Of course, we can replace the word 'row' in this definition by 'column' and in the same way define the *column rank* of M. For example, the first two rows of BA are independent as neither is a multiple of the other, but, as we have seen, the three rows do not form an independent set, as $\mathbf{r}_1$ is a linear combination of $\mathbf{r}_2$ and $\mathbf{r}_3$. Hence the row rank of BA is 2.

A remarkable theorem says that the row rank and column rank are always equal, and this common value is naturally called the *rank* of the matrix. What is more, you cannot create independence from dependence, by which is meant that the rank of a matrix can never be increased by taking its product with another matrix, or to put it in symbols,

$$\text{rank}(AB) \leq \min(\text{rank}(A), \text{rank}(B)). \tag{31}$$

In particular, since our 3×3 matrix can be factorized as a product of matrices that have rank at most 2 (because the first only has two columns and the second just two rows), it follows that BA must also have rank no more than 2. Indeed, as we have already noted, rank(BA) is 2.

Before closing this section, however, we hasten to point out that matrices do obey other laws of algebra. Importantly, multiplication is associative and is distributive over addition, so that for any three matrices A, B, and C we have that whenever one side of each of the following equations exists, then so does the other and they agree:

$$A(BC) = (AB)C \quad \text{and} \quad A(B+C) = AB + AC.$$

Each of these laws is a consequence of the corresponding law holding over the field of real numbers. The distributive law is easily checked, but associativity of multiplication is less obvious. Careful calculation of the entry in a typical position in each product reveals, however, that the product ABC is independent of how you bracket it, and this is ultimately a consequence of associativity of ordinary multiplication.

Networks

One of the most direct and fruitful lines of application of matrices and their multiplication is to the representation of networks. For

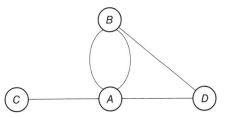

13. Network of cities and airline routes.

example, consider the network in Figure 13, where the four nodes represent cities A, B, C, D and we have drawn an edge joining two nodes if an airline flies on that route. Note that two airlines fly between cities A and B, each represented by an edge in the network.

It is possible to distil all the information of a network N into what is called the *incidence matrix*, M, of N. We simply number the nodes in some way; here we follow alphabetical order, and write down what is in this case a 4×4 matrix where the entry at position (i, j) is the number of edges connecting node i to node j. Look at what happens when we write down M and calculate its square:

$$M^2 = \begin{bmatrix} 0 & 2 & 1 & 1 \\ 2 & 0 & 0 & 1 \\ 1 & 0 & 0 & 0 \\ 1 & 1 & 0 & 0 \end{bmatrix} \begin{bmatrix} 0 & 2 & 1 & 1 \\ 2 & 0 & 0 & 1 \\ 1 & 0 & 0 & 0 \\ 1 & 1 & 0 & 0 \end{bmatrix} = \begin{bmatrix} 6 & 1 & 0 & 2 \\ 1 & 5 & 2 & 2 \\ 0 & 2 & 1 & 1 \\ 2 & 2 & 1 & 2 \end{bmatrix}.$$

The entries of the matrix product, M^2, tell you how many ways there are of travelling from one city to another via a third. For example, there are two ways of getting from D to B (via A). There are five return trips to B (one through D, but there are $2 \times 2 = 4$ via A as you have two choices of airline for each leg of the journey). In much the same way M^3, M^4, and, generally, M^n, are matrices whose entry at position (i, j) tells you the number of paths of length n from i to j. Since we are assuming that our edges allow passage equally in both directions, incidence matrices like M and its powers are *symmetric*, meaning that the entries at (i, j) and (j, i) always agree. But why does matrix multiplication automatically keep track of all this?

Ask yourself how we would count the total number of ways of going from a node i to a node j in a network via any choice of a third node k. For any particular choice of the intermediate node k, we would count all the edges between i and k (which might be zero) and multiply that number by the number of edges from k to our target node, j. This tells you how many paths of length 2 there are from i to j via k. We would then sum these numbers over all possible values of k to find the total number of paths of length 2 from i to j. However, that is exactly the sum that you work out to calculate the entry at (i, j) in the matrix product M^2 when you form the product of row i with column j.

Incidence matrices of various types are used to capture information about networks, and algebraic features of these matrices then correspond to and allow calculation of special features of the network. Indeed, network theory is one of the major applications of *linear algebra*, which is the branch of the subject that is largely represented by matrices and matrix calculations.

For example, we next explain the *Kirchhoff matrix K* of a network N. First, we define the *degree* of a node A in a network N to be the number of edges attached to A. In our network of Figure 13, for example, A has degree 4. In fact, the degree of the ith node must equal the sum of the ith row of M (and equally, by symmetry, the sum of the ith column). For a network N with n nodes, let D be the $n \times n$ matrix consisting entirely of zero entries except down the *principal diagonal*, which is the diagonal of the matrix from top left to bottom right, with the diagonal entry at (i, i) being the degree of the ith node. For the network N of our running example, we may write $D = \text{diag}(4, 3, 1, 2)$, where the notation is self-explanatory. Quite generally, we call a square matrix D a *diagonal matrix* if its non-zero entries lie exclusively on the principal diagonal. We now define the Kirchhoff matrix K of N to be $D - M$. In our example, this gives

$$K = D - M = \begin{bmatrix} 4 & 0 & 0 & 0 \\ 0 & 3 & 0 & 0 \\ 0 & 0 & 1 & 0 \\ 0 & 0 & 0 & 2 \end{bmatrix} - \begin{bmatrix} 0 & 2 & 1 & 1 \\ 2 & 0 & 0 & 1 \\ 1 & 0 & 0 & 0 \\ 1 & 1 & 0 & 0 \end{bmatrix}$$

$$= \begin{bmatrix} 4 & -2 & -1 & -1 \\ -2 & 3 & 0 & -1 \\ -1 & 0 & 1 & 0 \\ -1 & -1 & 0 & 2 \end{bmatrix}.$$

Note that in any Kirchhoff matrix, the sum of the entries in each row, and by symmetry in each column, is 0.

Through using K it is possible to compute the number of so-called *spanning trees* in the network N. A *tree* is a network that contains no cycles but in which there is a path between any two nodes: equivalently, a tree is a network in which there is a *unique* path between any two nodes. A *spanning tree* of a network is a tree within the network that contains all of the nodes. For example, the path C–A–D–B represents one spanning tree in N, and the reader should be able to find four others. The matrix K allows you to count the number of spanning trees in N without having to write down all the trees explicitly. To explain how K does this requires the notion of the *determinant* of a matrix, which is the subject of Chapter 9, where we will revisit this problem.

Geometric applications

Another application of matrices is to the geometry of transformations. We may use any given 2×2 matrix A as a means to create a mapping of points $P(x, y)$ in the Cartesian plane by writing the coordinates of P as a column vector, known in this context as a *position vector*, and multiplying that vector on the left by A to give a new point. For example,

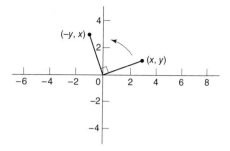

14. Geometric action of a matrix transformation.

$$\begin{bmatrix} 0 & -1 \\ 1 & 0 \end{bmatrix} \begin{bmatrix} x \\ y \end{bmatrix} = \begin{bmatrix} -y \\ x \end{bmatrix}.$$

In this instance, the matrix rotates each point (x, y) through 90° anticlockwise about the origin to the point $(-y, x)$ (Figure 14).

The *transformations* of the plane, also known as *mappings*, which we can create in this way have very special properties that are a consequence of the algebraic laws that matrices obey.

We first describe these transformations in general without reference to matrices. For convenience, we write the column vector $\begin{bmatrix} x \\ y \end{bmatrix}$ as **x** and, in general, typical column vectors will be denoted by bold letters. A transformation L of the points of the plane is called *linear* if for any two (column) vectors **x** and **y** and for each scalar (i.e. real number) r, the following two rules hold: $L(\mathbf{x} + \mathbf{y}) = L(\mathbf{x}) + L(\mathbf{y})$ and $L(r\mathbf{x}) = rL(\mathbf{x})$. Indeed, these rules can be combined into one single rule, that being that for any two vectors **x** and **y** and any two scalars r and s,

$$L(r\mathbf{x}+s\mathbf{y}) = rL(\mathbf{x}) + sL(\mathbf{y}). \qquad (32)$$

If we let L stand for a 2×2 matrix, these rules are satisfied, so that multiplication by such a matrix furnishes a source of linear transformations. More surprisingly, the converse is true: any linear transformation can be realized as multiplication by a suitable matrix, for we proceed as follows. The action of a linear transformation is determined by its action on the two *basis* vectors

$$\mathbf{b}_1 = \begin{bmatrix} 1 \\ 0 \end{bmatrix} \text{ and } \mathbf{b}_2 = \begin{bmatrix} 0 \\ 1 \end{bmatrix},$$

for let us suppose that

$$L\begin{bmatrix} 1 \\ 0 \end{bmatrix} = \begin{bmatrix} a \\ b \end{bmatrix} \text{ and } L\begin{bmatrix} 0 \\ 1 \end{bmatrix} = \begin{bmatrix} c \\ d \end{bmatrix}.$$

Then, since L is linear, we obtain the following through using (32), with x and y now denoting the multiplying scalars:

$$L\begin{bmatrix} x \\ y \end{bmatrix} = L\left(x\begin{bmatrix} 1 \\ 0 \end{bmatrix} + y\begin{bmatrix} 0 \\ 1 \end{bmatrix}\right) = xL\begin{bmatrix} 1 \\ 0 \end{bmatrix} + yL\begin{bmatrix} 0 \\ 1 \end{bmatrix} = x\begin{bmatrix} a \\ b \end{bmatrix} + y\begin{bmatrix} c \\ d \end{bmatrix}$$

$$= \begin{bmatrix} ax \\ bx \end{bmatrix} + \begin{bmatrix} cy \\ dy \end{bmatrix} = \begin{bmatrix} ax + cy \\ bx + dy \end{bmatrix} = \begin{bmatrix} a & c \\ b & d \end{bmatrix}\begin{bmatrix} x \\ y \end{bmatrix}.$$

Hence *any* linear transformation of the plane can be realized by multiplication by a matrix A, the columns of which are the images of the two basis vectors $\mathbf{b}_1$ and $\mathbf{b}_2$ under L, written in that order.

What is interesting is that if we *compose* linear mappings, which is to say act with two or more in succession, the outcome is again a linear mapping. This can be verified by working with the abstract definition (32), or we can see it by noting that the composition of two (or more) linear mappings can be realized by multiplying by each of the matrices A and B which represent each linear transformation, one after the other. The bonus here is that, by associativity of matrix multiplication, this composition can be realized by taking the product of the matrices, as $B(A\mathbf{x}) = (BA)\mathbf{x}$. The upshot of this is that, no matter how many linear transformations we act with, we can find a *single* matrix that has the effect of acting with each of these transformations in turn, in the order that we specify, by taking the corresponding product of matrices.

Two ways in which linear transformations come about are by rotations of the plane about the origin through a fixed angle, and by reflection in a line through the origin. For example, let us find the matrix that has the effect of the following three actions. For each point $P(x, y)$ in the plane, reflect P in the line $y = -x$, then rotate the resulting point through 90° clockwise about the origin, and finally reflect that point in the x-axis. To find the single matrix that does all this, we need to take the product of the matrices for each of these three transformations, with the first of them on the right and the final one on the left. The three matrices and their product are

$$\begin{bmatrix} 1 & 0 \\ 0 & -1 \end{bmatrix} \begin{bmatrix} 0 & 1 \\ -1 & 0 \end{bmatrix} \begin{bmatrix} 0 & -1 \\ -1 & 0 \end{bmatrix} = \begin{bmatrix} 0 & 1 \\ 1 & 0 \end{bmatrix} \begin{bmatrix} 0 & -1 \\ -1 & 0 \end{bmatrix} = \begin{bmatrix} -1 & 0 \\ 0 & -1 \end{bmatrix}$$

$$= -\begin{bmatrix} 1 & 0 \\ 0 & 1 \end{bmatrix},$$

where we have performed the matrix multiplications from left to right. This matrix has the effect that

$$\begin{bmatrix} x \\ y \end{bmatrix} \mapsto \begin{bmatrix} -x \\ -y \end{bmatrix},$$

which represents a half turn (180°) about the origin and so that is the net effect of the three transformations. (The symbol $\mapsto$ is read as 'maps to'.)

An important insight stems from how the general linear transformation represented by the matrix

$$A = \begin{bmatrix} a & b \\ c & d \end{bmatrix}$$

acts on the unit square with corners at the origin, $O = (0, 0)$, and the points $(1, 0)$, $(0, 1)$, and $(1, 1)$. The origin remains fixed while the other three vertices are sent to (a, c), (b, d), and $(a + b, c + d)$, respectively (Figure 15).

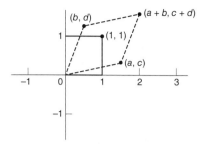

15. Matrix acting on the unit square.

The unit square is then mapped to this parallelogram, the area of which can be shown to equal $|ad - bc|$. The quantity $\Delta = ad - bc$ is known as the *determinant* of A and it tells us by how much the area of any figure in the plane is magnified (or contracted) when acted on by A. If $\Delta < 0$ then A also reverses the orientation of the figure, in that if a triangle has vertices P, Q, and R, in that order when traced in the anticlockwise direction, then the image triangle will have the respective vertices P', Q', and R' oriented clockwise: it will not be possible to place the original triangle onto its image without breaking out of the plane and flipping it over.

For example, consider our matrices

$$R = \begin{bmatrix} 0 & -1 \\ 1 & 0 \end{bmatrix}, \quad S = \begin{bmatrix} 1 & 0 \\ 0 & -1 \end{bmatrix},$$

so that R rotates each point through 90° anticlockwise about the origin while S reflects each point in the x-axis. The matrix R does not alter either the area or the orientation of a figure, and so its determinant is $\det(R) = (0 \times 0) - ((-1) \times 1) = 0 + 1 = 1$. On the other hand, although S does not change areas either, S does swap orientation and this is manifested in the value of its determinant: $\det(S) = 1 \times (-1) - 0 \times 0 = -1$.

The special case where $\Delta = 0$ is when $ad = bc$, which occurs if $a/c = b/d$. In this case the two position vectors from the origin that make up the columns of the matrix both point along the same line through the origin with a common gradient, c/a, and the image parallelogram degenerates to a line segment. The image of any shape is then collapsed along this line and so the area of that image is 0. It remains the case, however, that the absolute value of the determinant of A, $|\Delta|$, is the multiplier of area under the transformation defined by A.

This meeting of the algebraic and geometric worlds is important and we can already see an interesting theorem on offer. If we act with two linear transformations in turn, represented by matrices A and then B, then areas will be multiplied by the multiplier of A followed by that of B, which of course must be the multiplier of the product matrix BA. Moreover, the sign of $\det(BA)$ will be the product of the signs of $\det(A)$ and $\det(B)$, as BA will preserve orientation if both mappings preserve or both reverse orientation but not otherwise. The upshot of this is that the determinant of a product of two square matrices equals the product of their determinants, and since $\det(A)\det(B) = \det(B)\det(A)$ we may write

$$\det(AB) = \det(A)\det(B) = \det(BA).$$

From an algebraic viewpoint, this may seem surprising. It is true nonetheless, extends to higher dimensions, and can be explained in algebraic terms. It is the geometric interpretation, however, that renders the result transparent and, we might even say, 'obvious'.

Chapter 8
Matrices and groups

Inverses and groups

Curiosity demands that we persevere with our example of geometric matrix products to see what happens when we compose the mappings involved. Specifically, let R now denote the transformation that rotates each point about the origin through one quarter turn ($90°$) anticlockwise, and let S denote reflection in the x-axis. Using our matrix representation, we then have

$$R = \begin{bmatrix} 0 & -1 \\ 1 & 0 \end{bmatrix}, \quad S = \begin{bmatrix} 1 & 0 \\ 0 & -1 \end{bmatrix}.$$

We look to find the transformations that can be *generated* by R and S, meaning the mappings that we can produce by composing R and S with each other, in any order, and any number of times.

First it is clear that, since S is a reflection, if we act with S and act with S again (we write this as S^2), the net effect is to return every point to its original position. In terms of matrices, we have

$$S^2 = \begin{bmatrix} 1 & 0 \\ 0 & -1 \end{bmatrix} \begin{bmatrix} 1 & 0 \\ 0 & -1 \end{bmatrix} = \begin{bmatrix} 1 & 0 \\ 0 & 1 \end{bmatrix},$$

and this latter matrix is called the *identity matrix*, denoted by I. The effect of multiplying I by any column vector is to return the same column vector, and so we say that I represents the *identity transformation* that leaves every point fixed. More generally, we speak of the $n \times n$ matrix $I = I_n$, which is the diagonal matrix with a principal diagonal that consists entirely of 1's. For any matrix A of a size that allows matrix multiplication, we always have $AI = A = IA$, so that I behaves with respect to matrix multiplication like the multiplicative identity 1 does with respect to ordinary multiplication.

In symbols, we have that $S^2 = I$ and we say that S is its own *inverse* matrix. In general, for a square matrix A, we say that another square matrix B is the *inverse* of A if $AB = BA = I$ and we write $B = A^{-1}$ for, it exists, the inverse of A is certainly unique. To see this, suppose that we had two inverses, B and C, of A, so that $BA = CA = I$. By multiplying this equation on the right by B we then get

$$B(AB) = C(AB) \Rightarrow BI = CI \quad \text{and so} \quad B = C.$$

It follows from the symmetry of the definition that A is equally the inverse of B, and so $A = B^{-1}$ also.

However, not every square matrix has an inverse: for example, a square *zero matrix* Z, which is a matrix where all entries are 0, cannot have an inverse, as for any matrix A of the same size we have $ZA = AZ = Z$: Z is the matrix analogue of the number 0. However, there are also other *singular* matrices, which mean matrices without inverses, and we shall have more to say about this.

We can see directly, however, that our matrix R does have an inverse. We only need ask ourselves, how can we undo the effect of acting with R? Since R represents an anticlockwise quarter turn, we just need the matrix that turns the plane through a quarter turn in the clockwise direction. We can also achieve this by turning through three quarters of a full turn in the anticlockwise direction. The reason to draw attention to this alternative is that this latter transformation will come from acting with R three times or, in terms of our new notation, $R^{-1} = R^3$:

$$R^{-1} = R^3 = \begin{bmatrix} 0 & 1 \\ -1 & 0 \end{bmatrix}.$$

We have thus discovered that while there are just two distinct powers of S, there are four for our rotation R, giving us a group of transformations (for it will transpire that it is a group), that being $H = \{I, R, R^2, R^3\}$.

Recall from Chapter 6 that a *group* is a set, together with an associative binary operation, which possesses an identity element with respect to which each member of the group has an inverse. In the case of H, the operation is composition of transformations or, equivalently, multiplication of the corresponding matrices. Matrix multiplication is indeed associative, the identity element I is represented by the identity matrix, and, crucially, each member A of the set possesses an inverse, A^{-1}, as we shall now check.

In the case of H, I is self-inverse (which is always the case), as is R^2 for we have $R^2 \cdot R^2 = R^4 = I$, and we have already noted that $R \cdot R^3 = I$, showing that R and R^3 are mutual inverses. Therefore H is an example of a group. However, H is simply a 4-cycle meaning that H just consists of four powers of R, with $R^5 = R$, $R^6 = R^2$, and so on, and therefore H is a copy of $\mathbb{Z}_4$, the group of integers modulo 4. Nonetheless, H is a part of (we say H is a *subgroup* of) a larger eight-element group, D. The other four members of D can be generated by taking the product of each member of H with S, which gives $\{S, RS, R^2S, R^3S\}$. By calculating the matrices representing each of these four mappings or by considering their geometric actions, we can see that each of these products represents a reflection in a line through the origin. Specifically, S, RS, R^2S, R^3S correspond to reflections in the lines $y = 0$, $y = x$, $x = 0$, and $y = -x$ respectively, these four reflections being self-inverse mappings. (Rotating each of these lines by $45°$ anticlockwise takes you from one to the next.)

Table 1. Multiplication of the group of symmetries generated by R and S.

	I	R	R^2	R^3	S	RS	R^2S	R^3S
I	I	R	R^2	R^3	S	RS	R^2S	R^3S
R	R	R^2	R^3	I	RS	R^2S	R^3S	S
R^2	R^2	R^3	I	R	R^2S	R^3S	S	RS
R^3	R^3	I	R	R^2	R^3S	S	RS	R^2S
S	S	R^3S	R^2S	RS	I	R^3	R^2	R
RS	RS	S	R^3S	R^2S	R	I	R^3	R^2
R^2S	R^2S	RS	S	R^3S	R^2	R	I	R^3
R^3S	R^3S	R^2S	RS	S	R^3	R^2	R	I

However, no more than these eight mappings may be obtained, so it follows that D forms a group of transformations, with each of the four reflections being self-inverse. The table of the group operation is displayed in Table 1, where the entry at position (i, j) in the body of the table is the product $\alpha\beta$ of the matrix α in row i times the matrix β in column j.

Any product of any length is equal to one of these eight mappings. Indeed any product, however long, can be simplified to one of the eight given forms by using just the rules $R^4 = S^2 = I$ and $SR = R^3S$. Group theorists would say that the group D is *presented by the generators* R and S, subject to these three *relations*. For example, making free use of these equations, we get

$$SR^2S = (SR)RS = (R^3S)RS = R^3(SR)S = R^3(R^3S)S$$

$$= (R^4)R^2(S^2) = R^2.$$

A key observation about D, known as the *dihedral group*, is that its operation is *not* commutative: $SR = R^3S \neq RS$. A more general observation that applies to every group is that its table of products displays the *Latin square property*, meaning that every row and every column features each member of the group exactly once. These facts help us complete the table.

The cyclic group H has been seen before in another guise, for the powers of the imaginary unit i are four in number, i.e. $G = \{1, i, -1, -i\}$, and correspond to the list $H = \{1, R, R^2, R^3 = R^{-1}\}$ in that order: we say that the two groups G and H are *isomorphic* and that the correspondence $1 \mapsto 1$, $R \mapsto i$, $R^2 \mapsto -1$, $R^{-1} \mapsto -i$ is an *isomorphism between G and H*, meaning that their tables of products are identical under this renaming of elements.

The purpose of the word 'isomorphism' is to convey the fact that although the two groups are different, the first being a collection of complex numbers under multiplication while the second is a group of rotations under the operation of composition, the two groups are essentially the same as regards their algebraic structure. This is not a mere coincidence, as multiplication by the imaginary unit i acts to rotate each number in the complex plane through a quarter of a turn about the origin. To see this we need only note that for any complex number $z = x + iy$, we have $iz = xi + i^2 y = -y + ix$: in terms of points in the Argand plane, $(x, y) \mapsto (-y, x)$, and the latter is the point we arrive at by turning (x, y) through $90°$ anticlockwise about the origin. This link lets us identify multiplication by i with a turn of one right angle in the plane, thereby reaffirming the fundamental nature of the imaginary unit.

Matrix inverses

If a collection of matrices represent the elements of a group, such as the eight matrices that represent the dihedral group D, then each of these matrices A will have an inverse A^{-1}, such that $AA^{-1} = A^{-1}A = I$, the identity matrix. This prompts the twin questions of when the inverse of a square matrix A exists and, if it does, how to find it. First we make a few simple observations about inverses in general.

A very useful rule is that the inverse of a product of two square matrices is the product of the inverses in inverse order, which is to say that $(AB)^{-1} = B^{-1}A^{-1}$, as is verified at once from the definition:

$$(AB)(B^{-1}A^{-1}) = A(BB^{-1})A^{-1} = AIA^{-1} = AA^{-1} = I. \qquad (33)$$

It is worth noting that the manipulations represented by (33) hold in any group: the inverse of a product is the product of the inverses with the order reversed. It is important to respect this reversal of order, as for matrix multiplication, and for group products in general, order matters.

There is an important but simple operation akin to taking the inverse that applies to any matrix A, which is that of taking its *transpose*, A^T: the rows of A^T written from left to right are the columns of A written from top to bottom. (It is sometimes convenient to write a column vector using the transpose symbol, for example $(2, -1, 4)^T$). As an example,

$$A = \begin{bmatrix} 2 & -1 \\ 7 & 4 \\ 0 & 3 \end{bmatrix} \Leftrightarrow A^T = \begin{bmatrix} 2 & 7 & 0 \\ -1 & 4 & 3 \end{bmatrix}. \qquad (34)$$

We are entitled to use the double arrow '$\Leftrightarrow$' in (34) as, just like for a square matrix A it is the case that $(A^{-1})^{-1} = A$, it is true quite generally that $(A^T)^T = A$. What is a little less obvious is that the taking of the transpose also obeys the rule $(AB)^T = B^T A^T$. As we have mentioned previously in connection with incidence matrices of networks, a matrix that equals its own transpose is called *symmetric*: a symmetric matrix is necessarily square and is one that satisfies $a_{ij} = a_{ji}$ throughout.

Suppose now that we have n equations in n unknowns, $x_1, x_2, \ldots, x_n$. We may represent this system by a single matrix equation, $A\mathbf{x} = \mathbf{b}$, where $\mathbf{x}$ represents the column of unknowns, $\mathbf{b}$ represents the RHS of the equation set, and A is the $n \times n$ coefficient matrix. If we already possessed the inverse, A^{-1}, we could solve this system immediately by multiplying both sides of the equation on the left by A^{-1}, for that gives $A^{-1}A\mathbf{x} = A^{-1}\mathbf{b} \Rightarrow \mathbf{x} = A^{-1}\mathbf{b}$. On the other hand, we have the elimination technique to solve such a system, which suggests that we might seek to use that to find how to invert our matrix A.

The operations deployed in the elimination method used to solve a set of simultaneous linear equations affect the coefficient matrix in one of three ways:

1. Multiply a row by a non-zero constant a.
2. Interchange two rows.
3. Add a multiple a of one row to another.

Each of these *row operations*, as they are called, can be effected by multiplying on the left by a suitable matrix. The following examples illustrate this:

$$B = \begin{bmatrix} 1 & 0 \\ 0 & a \end{bmatrix}, \quad C = \begin{bmatrix} 0 & 1 \\ 1 & 0 \end{bmatrix}, \quad D = \begin{bmatrix} 1 & a \\ 0 & 1 \end{bmatrix}. \tag{35}$$

For any 2×2 matrix A, the product BA outputs A with its second row multiplied by a, CA is the same as A but with the rows interchanged, and multiplying A on the left by D adds a times the second row of A to the first row of A. Specifically, for this last example,

$$\begin{bmatrix} 1 & a \\ 0 & 1 \end{bmatrix} \begin{bmatrix} u & v \\ r & s \end{bmatrix} = \begin{bmatrix} u+ar & v+as \\ r & s \end{bmatrix} = \begin{bmatrix} u & v \\ r & s \end{bmatrix} + a \begin{bmatrix} r & s \\ 0 & 0 \end{bmatrix}.$$

In general, we get the row matrix that effects a particular row operation by applying that row operation to the identity matrix I. Although there is no need to introduce these *elementary row matrices* in any practical calculation, there is an important consequence to their being there. Suppose that we reduce A to I by a sequence of row operations that have row matrices $M_1, M_2, \ldots, M_{m-1}, M_m$, say, so that we have $(M_m M_{m-1} \ldots M_2 M_1)A = I$. It follows that the product of the matrices preceding A must equal A^{-1}. We can, however, find A^{-1} from this analysis without writing down the matrices M_i explicitly. The scheme is often expressed in the following fashion:

$$[A|I] \to [I|A^{-1}],$$

where the arrow represents row reduction of operations applied to A to reduce it to I. The $n \times 2n$ matrix on the left is known as the *augmented matrix* of A. Since the row operations used are executed on both the left-hand portion, A, and the right-hand portion, I, of the augmented matrix, the net effect on I will be to multiply it too by A^{-1}, so that the right half of the augmented matrix is indeed transformed into $A^{-1}I = A^{-1}$, as the scheme suggests.

To illustrate this method, we find the inverse of the 3×3 matrix forming the left-hand portion of the augmented matrix below, so that $[A|I]$ in this case is given by

$$\begin{bmatrix} 1 & 1 & 1 & | & 1 & 0 & 0 \\ 3 & 4 & 5 & | & 0 & 1 & 0 \\ 3 & 6 & 10 & | & 0 & 0 & 1 \end{bmatrix}.$$

We begin by *pivoting* on the top left-hand corner, meaning that we clear all other entries in the first column by means of the row operations $R_2 \mapsto R_2 - 3R_1$ (meaning that row 2 is replaced by row 2 minus 3 row 1) and $R_3 \mapsto R_3 - 3R_1$ to give

$$\begin{bmatrix} 1 & 1 & 1 & | & 1 & 0 & 0 \\ 0 & 1 & 2 & | & -3 & 1 & 0 \\ 0 & 3 & 7 & | & -3 & 0 & 1 \end{bmatrix} \to \begin{bmatrix} 1 & 0 & -1 & | & 4 & -1 & 0 \\ 0 & 1 & 2 & | & -3 & 1 & 0 \\ 0 & 0 & 1 & | & 6 & -3 & 1 \end{bmatrix},$$

where we pass to the second matrix by pivoting on the second diagonal entry via the operations $R_1 \mapsto R_1 - R_2$ and

$R_3 \mapsto R_3 - 3R_2$. Note that this does not alter any entry in the first column, as only the pivotal entry of the first column is not zero—in this way, the process moves forward and we do not resurrect non-zero entries where they are unwanted. Finally, we clear above the third diagonal pivot to reduce the left-hand portion of the matrix to I_3 using the operations $R_1 \mapsto R_1 + R_3$ and $R_2 \mapsto R_2 - 2R_3$, to obtain

$$\begin{bmatrix} 1 & 0 & 0 & | & 10 & -4 & 1 \\ 0 & 1 & 0 & | & -15 & 7 & -2 \\ 0 & 0 & 1 & | & 6 & -3 & 1 \end{bmatrix} \Rightarrow A^{-1} = \begin{bmatrix} 10 & -4 & 1 \\ -15 & 7 & -2 \\ 6 & -3 & 1 \end{bmatrix}.$$

It may be readily checked that $AA^{-1} = I$, and readers are now in a position to practise some examples of their own, but be warned: the previous example is especially nice—it is rare for a matrix and its inverse to consist entirely of integers. During your calculation, you may sometimes avoid fractions by pivoting on diagonal entries that are not 1, but, of course, if the answer has fractions in it they are bound to make their appearance sooner or later. An *integer matrix* (one with all integer entries) has an integer matrix inverse if and only if its *determinant* is ±1: we shall justify this claim in Chapter 9. However, necessary and sufficient conditions for a square matrix A to possess an inverse may be given just in terms of the notion of rank, which we introduced in Chapter 7, and we can now outline how this comes about.

A square matrix A is invertible if and only if A is of full rank. If A is not of full rank, we can use the dependency within the rows to reduce A by row operations to a matrix C with a row of zeros, and such a matrix must be singular, as that row of zeros will persist in any product of the form CB and so cannot give the identity matrix. Since each row operation is invertible, it can be inferred that A too has no inverse and so only matrices of full rank are invertible.

Conversely, any square matrix A of full rank can be inverted. This is because in that case the reduction process that takes $[A|I]$ to $[I|A^{-1}]$ will never get stuck—the fact that A is of full rank ensures that a zero that arises in a pivot position can be exchanged for a non-zero entry and the process will continue, eventually yielding the inverse of A in the reduced augmented matrix $[I|A^{-1}]$. Therefore, provided that A is of full rank, its inverse exists and may be found in this way. In the next chapter we shall see that another characterization of invertibility is possible in terms of a single number associated with the matrix, its determinant.

Chapter 9
Determinants and matrices

Determinants

To further address the general question of matrix invertibility, we begin with an analysis of the case of a 2×2 matrix A. Supposing that A may be inverted, it follows that A has no column of zeros, so that, by swapping the first and second rows if necessary, we may assume that the entry a in the top left-hand corner is not 0, in which case we can clear below it and begin the process of creating the inverse by the operation $R_2 \mapsto R_2 - (c/a)R_1$:

$$\begin{bmatrix} a & b & 1 & 0 \\ c & d & 0 & 1 \end{bmatrix} \to \begin{bmatrix} a & b & 1 & 0 \\ 0 & d - cb/a & -c/a & 1 \end{bmatrix} = \begin{bmatrix} a & b & 1 & 0 \\ 0 & (ad - bc)/a & -c/a & 1 \end{bmatrix}.$$

To proceed further, we will seek to clear the entry b at position (1, 2) down to 0 by a suitable subtraction of the second row from the first, and this will involve the reciprocal of the entry now at position (2, 2) of our matrix. In particular, we will need to divide by the quantity $\Delta = ad - bc$, which we recognize as the *determinant* of A that we met in Chapter 7. Clearly, we therefore require that $\Delta \neq 0$ to proceed to this stage, for otherwise the matrix we are trying to invert will have a zero second row. It is worth noting that the condition $\Delta \neq 0$ includes the one we have already noted, that at least one of a and c is non-zero, for otherwise $ad - bc = 0 - 0 = 0$.

Assuming that the determinant is not 0, we may get the diagonal entries reduced to 1 via the operations $R_1 \mapsto (1/a)R_1$ and $R_2 \mapsto (a/\Delta)R_2$; finally, we render the left portion of the augmented matrix as I_2 by the operation $R_1 \mapsto R_1 - (b/a)R_2$ to yield

$$\begin{bmatrix} 1 & b/a & 1/a & 0 \\ 0 & 1 & -c/\Delta & a/\Delta \end{bmatrix} \to \begin{bmatrix} 1 & 0 & 1/a + bc/a\Delta & -b/\Delta \\ 0 & 1 & -c/\Delta & a/\Delta \end{bmatrix}.$$

However, the term at position (1, 3) simplifies to

$$\frac{\Delta + bc}{a\Delta} = \frac{ad - bc + bc}{a\Delta} = \frac{ad}{a\Delta} = \frac{d}{\Delta}$$

and so, provided that $\Delta \neq 0$, we may take the common denominator Δ outside the matrix to reveal the inverse of A:

$$A^{-1} = \frac{1}{\Delta} \begin{bmatrix} d & -b \\ -c & a \end{bmatrix}.$$

And this is representative of the general situation: the augmented matrix method will yield the inverse, A^{-1}, of a square matrix A unless the determinant of A, often written as $\Delta = |A|$, is equal to 0, in which case no inverse exists. However, we have not yet explained what this number Δ is in the case of a general square matrix, and before we do we explore a little further.

In three dimensions, the absolute value of the determinant, $\det(A)$, of a linear transformation represented by a matrix A is the multiplier of volume. The columns of A are the images of the position vectors of the sides of the unit cube, $\mathbf{b}_1 = (1, 0, 0)^T$, $\mathbf{b}_2 = (0, 1, 0)^T$, $\mathbf{b}_3 = (0, 0, 1)^T$, and they define a three-dimensional version of a parallelogram, a *parallelepiped*, the volume of which is $|\det(A)|$, with the orientation of the figure being reversed if $\Delta < 0$.

But how do we define the determinant of a 3×3 matrix? First we shall explain how it is calculated and then return to the question as to just how it is defined. The value of $\det(A)$ is found in a certain prescribed fashion based on any row or any column of the array. Choosing to work from the first row of the matrix, we multiply each entry by the 2 × 2 determinant of the four entries that remain when we cross out the row and column of that entry, and then sum these three numbers with alternating signs. (The sign associated with each position is as indicated in the alternating pattern in (36).) For example, using the alternative notation $|A|$ for $\det(A)$, we have

$$|A| = \begin{vmatrix} 1 & 2 & 3 \\ -1 & 0 & 2 \\ 2 & 1 & 2 \end{vmatrix}; \quad \begin{vmatrix} + & - & + \\ - & + & - \\ + & - & + \end{vmatrix} \quad (36)$$

$$= 1 \begin{vmatrix} 0 & 2 \\ 1 & 2 \end{vmatrix} - 2 \begin{vmatrix} -1 & 2 \\ 2 & 2 \end{vmatrix} + 3 \begin{vmatrix} -1 & 0 \\ 2 & 1 \end{vmatrix}$$

$$= 1(0 - 2) - 2(-2 - 4) + 3(-1 - 0) = -2 + 12 - 3 = 7.$$

To emphasize the point that we may use any row or column, we calculate this determinant again, this time using the second-column expansion:

$$-2 \begin{vmatrix} -1 & 2 \\ 2 & 2 \end{vmatrix} + 0 \begin{vmatrix} 1 & 3 \\ 2 & 2 \end{vmatrix} - \begin{vmatrix} 1 & 3 \\ -1 & 2 \end{vmatrix}.$$

$$= -2(-2 - 4) + 0 - (2 - (-3)) = 12 + 0 - 5 = 7.$$

What is more, the pattern of definition extends to 4 × 4 and, in general, $n \times n$ matrices A in that $|A|$ may be calculated using any row or column and taking the sum of each signed entry multiplied by the determinant of the *minor* $(n-1) \times (n-1)$ matrix that results from deleting the row and column of the entry.

The reader may like to try their hand on the 3 × 3 matrix (29) in Chapter 7. However, since that matrix is not of full rank, we know in advance that its determinant must be 0. A similar remark applies to the 4 × 4 Kirchhoff matrix K, also featured below. Since the sum of the rows of any Kirchhoff matrix is the zero vector, such a matrix K is not of full rank and therefore has a determinant of 0.

The reason why a determinant computation is independent of the line in the matrix on which the calculation is based looks very mysterious and so stands in need of some explanation. In each case the result of the computation turns out to be the following particular sum. The determinant Δ of an $n \times n$ matrix $A = (a_{ij})$ is a certain sum of signed products of the elements of A. The products concerned are all of length n and are formed by taking one member from each row and each column: there are $n!$ such products, for in forming them we have n choices from the first row, but only $n - 1$ from the second (as we cannot repeat a column), $n - 2$ from the third row, and so on.

The accompanying sign is determined as follows. Each product defines a permutation, which is to say a reordering, of the numbers 1 to n by which $i \mapsto j$ if a_{ij} is the entry in the product chosen from row i. The associated sign of the product is then + or − according as that permutation can be effected by an even or by an odd number of transpositions, a transposition being a permutation that simply swaps the order of two positions.

For the case of $n = 2$ with entries as above we have $2! = 2$ terms, which are ad and bc. The permutation associated with ad is then $1 \mapsto 1, 2 \mapsto 2$, as a is in position $(1, 1)$ and d in position $(2, 2)$. This is the identity permutation that requires zero transpositions, and so is even, and the sign accompanying the product ad is +. On the other hand, the permutation of the term bc is the transposition that swaps 1 and 2 (as b is at position $(1, 2)$, so that $1 \mapsto 2$, and c is at $(2, 1)$, so that $2 \mapsto 1$). Since the number of transpositions used is odd, the associated sign is −. This gives the expected result that $\Delta = ad - bc$.

For a general 3×3 determinant there are $3! = 6$ possible products, and we obtain

$$\begin{vmatrix} a & b & c \\ d & e & f \\ g & h & i \end{vmatrix} = aei - afh - bdi + bfg + cdh - ceg;$$

for instance, the permutation associated with the term bdi is
$1 \mapsto 2, 2 \mapsto 1$, and $3 \mapsto 3$, which corresponds to a single
transposition that swaps 1 and 2; this therefore has an odd
number of transpositions, so that bdi carries a minus sign.

As an application of determinants to networks, we return to the
Kirchhoff matrix K of the network N in the example in Chapter 7:

$$K = \begin{bmatrix} 4 & -2 & -1 & -1 \\ -2 & 3 & 0 & -1 \\ -1 & 0 & 1 & 0 \\ -1 & -1 & 0 & 2 \end{bmatrix}.$$

A Kirchhoff matrix has the special property that, up to sign, the
determinant of each *cofactor* matrix, which is a submatrix of K
that comes from deleting any row and any column, always has the
same value. What is more, that number tells you exactly how many
spanning trees are possessed by the associated network. For
example, after deleting the first column and bottom row, we are
left to compute:

$$\begin{vmatrix} -2 & -1 & -1 \\ 3 & 0 & -1 \\ 0 & 1 & 0 \end{vmatrix} = - \begin{vmatrix} -2 & -1 \\ 3 & -1 \end{vmatrix} = -((2 - (-3)) = -5,$$

where we have computed the 3×3 determinant by expansion of
the bottom row (because of the convenience granted by the two 0
entries). The number of spanning trees of our network N is then 5,
as mentioned in Chapter 7.

In general, the determinant of all the 3×3 cofactor matrices of K
will be ± 5, with the sign associated with each entry being opposite
to that of all of its neighbours, in accord with the pattern of signs
indicated in (36). Each 3×3 determinant multiplied by the
accompanying sign is known as a *cofactor* of the 4×4 matrix.
This result extends to square matrices of any size, and Kirchhoff's
Theorem is that, for a Kirchhoff matrix, each of the these cofactors
is the same and equals the number of spanning trees of the

associated network. Of course, if our original network is a tree, then the determinant of each of these cofactors is 1, and in this way we have a source of integer matrices with determinant 1.

Properties of determinants

The calculations involved in finding determinants can be long, but from a mathematical viewpoint the concept is best thought about through its principal properties, which may now be revealed. For example, since row and column expansions yield the same result, it follows that a square matrix A has the same determinant as its transpose, A^T.

The key properties of determinants, however, are intimately connected with the row operations introduced earlier to solve sets of equations. If we multiply a row (or column) by a constant a, then the determinant of the matrix is likewise multiplied by a because of the introduction of this common factor a into each of the terms in the sum. If, on the other hand, we swap two adjacent rows (or columns) of A, this swaps the sign of every term in the determinant sum, and so swaps the sign of $|A|$ overall. Since any two rows may be exchanged by swapping pairs of adjacent rows an odd number of times, it follows that switching any two rows of a square matrix changes the sign of the determinant. Finally, if we add a multiple of one row to another, the determinant remains the same. This is a crucial property and it stems from the more general fact that if we add an arbitrary row vector **r** to a row of the matrix A, then the determinant of the new matrix is the sum of $\det(A)$ and the determinant of the matrix with that row of A replaced by **r**. If, however, **r** is some multiple of a row already in the matrix A, this second matrix is not of full rank and so the second determinant is 0, which is why the operation of adding one row (or any multiple of that row) to another leaves the determinant unchanged.

A function of matrices that has the three properties of the previous paragraph is necessarily the determinant function $|A|$ or a multiple of it, such as $2|A|$. If we add the *normalizing* condition that $|I| = 1$, then the determinant is the one and only function that has all four properties.

Mathematicians are very fond of results of this kind because proofs often argue from the properties of the subject in question. If you have a concise list of properties that are not only valid but also capture the essence of the subject of your theorem, you should be able to prove your result using only those properties. This leads to good proofs that demonstrate clearly why the theorem in question holds.

For example, in Chapter 7 we gave an intuitive argument as to why $|AB| = |A| \times |B|$ by appeal to the fact the determinant represents the area multiplier, or in general the volume multiplier, of a linear transformation. The multiplicative property of the determinant can be explained algebraically, however, by three observations.

From rank considerations, both sides of the equation will be 0 if and only if at least one of matrices A or B is singular. Otherwise, we can factorize A as a product of elementary row matrices. However, by using the properties of the determinant that we have just listed, it is quite simple to show that for each elementary row matrix R, it is the case that $|RB| = |R| \times |B|$. The general result now follows from repeated application of this fact.

In consequence, we note that the determinant of the inverse of a matrix A is the inverse of its determinant, for we have

$$1 = |I| = |AA^{-1}| = |A| \times |A^{-1}| \Rightarrow |A^{-1}| = |A|^{-1}. \qquad (37)$$

In Chapter 8 we mentioned that a square integer matrix A has an integer matrix for its inverse if and only if $|A| = \pm 1$. In one direction, this follows from (37), for it is only possible for both $|A|$

and $|A|^{-1}$ to be integers if $|A| = |A|^{-1} = \pm 1$. That the converse is true follows from applying the following compact formula for the solution of a square $n \times n$ equation system $A\mathbf{x} = \mathbf{b}$, that being

$$x_i = \frac{\det(A_i)}{\det(A)},$$

where x_i is the ith unknown in $\mathbf{x}$ and A_i is the matrix that results when the ith column of A is replaced by $\mathbf{b}$.

This rule is named after the French mathematician Gabriel Cramer (1704–52). In particular, taking $\mathbf{b}$ to be each of the columns of the identity matrix in turn, the corresponding solution vectors $\mathbf{x}$ provide a list of the columns of A^{-1}, and in this way we obtain a method for finding inverses through determinants, which also provides some nice theoretical results as well. In particular, since the determinant of a square integer matrix A is clearly an integer, it follows from Cramer's Rule that if $\det(A) = \pm 1$ and A is an integer matrix, then so is its inverse, A^{-1}.

Interestingly, in 2011 Habgood and Arel proved that the number of calculations required to implement Cramer's Rule is of the same order as that for the standard Jordan–Gaussian elimination method. The number of basic arithmetic operations involved in both approaches is of the order n^3. In other words Cramer's Rule is not just a neat formula; instead, despite involving determinants, the rule can be applied fast enough to solve large equation systems.

Eigenvalues and eigenvectors

In Chapter 7 we worked with geometric examples in the plane, but the same ideas apply in higher dimensions. Let us take an example of a linear mapping, L, in three dimensions through using the basis vectors $\mathbf{b}_1 = (1, 0, 0)^T$, $\mathbf{b}_2 = (0, 1, 0)^T$, and $\mathbf{b}_3 = (0, 0, 1)^T$. Our linear mapping L permutes the basis vectors in cyclic order, meaning that $\mathbf{b}_1 \mapsto \mathbf{b}_2$, $\mathbf{b}_2 \mapsto \mathbf{b}_3$ and $\mathbf{b}_3 \mapsto \mathbf{b}_1$, and so has as its matrix

$$A = \begin{bmatrix} 0 & 0 & 1 \\ 1 & 0 & 0 \\ 0 & 1 & 0 \end{bmatrix}.$$

Each of the basis vectors is mapped onto another by rotating about the line $x = y = z$, and since three applications of A map each basis vector back onto itself, that angle of rotation is $360°/3 = 120°$. Moreover, this observation amounts to saying that $A^3 = I$, so that $A^{-1} = A^2$. This can be seen geometrically as well in that A^{-1} will effect a rotation of $120°$ in the opposite sense, which is the same as a rotation of $240°$ in the original direction, which is of course the effect of the matrix A^2.

This example serves to introduce the topic of this section, for consider any point on the axis of rotation with position vector **x**. Since this point will remain fixed under this rotation, without calculation we infer that $A\mathbf{x} = \mathbf{x}$. We can easily check this directly: the vector **x** has the form $(a, a, a)^T$ and so the previous equation may be verified at once.

In general, for any square matrix A, we say that a non-zero vector **x** is an *eigenvector* of A with *eigenvalue* λ if $A\mathbf{x} = \lambda\mathbf{x}$. In our example, the vector $(1, 1, 1)$ is an eigenvector with eigenvalue 1. (It is easy to see from the definition that any multiple of an eigenvector **x** will also satisfy the defining equation, a point worth noting from the beginning.) The notion of eigenvector is fundamental throughout all of linear mathematics, as the behaviour of transformations can often be explained very clearly once these directions and the corresponding eigenvalues are known.

However, not all matrices have eigenvectors, for, by definition, an eigenvector **x** has to be mapped by A onto a multiple $\lambda\mathbf{x}$ of itself, which is to say that **x** and $A\mathbf{x}$ point in the same or in opposing directions, depending on the sign of the eigenvalue λ. For instance, a rotation *in the plane* about the origin through an angle θ, with

$0 < \theta < 180°$, changes the direction of every vector—no eigenvectors can be found here.

A property of note is that the eigenvalues of A^{-1} are just the inverses of the eigenvalues of A, and A and A^{-1} share the corresponding eigenvectors too, for if **x** is an eigenvector of A with eigenvalue $\lambda \neq 0$ then

$$A^{-1}\mathbf{x} = A^{-1}(\lambda^{-1}\lambda\mathbf{x}) = \lambda^{-1}A^{-1}(\lambda\mathbf{x}) = \lambda^{-1}A^{-1}(A\mathbf{x}) = \lambda^{-1}\mathbf{x}.$$

This last equation also alerts us to the fact that there is a relationship between invertibility of a matrix and its eigenvalues, which turns out to be particularly important. Although we exclude the zero vector as an eigenvector (as $A\mathbf{0} = \lambda\mathbf{0} = \mathbf{0}$ for every matrix A and every real number λ, so this tells us nothing in particular about A), we do not prohibit the number 0 from being an eigenvalue. If this were the case, we would have some *non-zero* vector **x** such that $A\mathbf{x} = 0\mathbf{x} = \mathbf{0}$. However, if we could multiply both sides of this equation on the left by A^{-1}, we would get $A^{-1}A\mathbf{x} = \mathbf{0}$, which gives the contradiction that $\mathbf{x} = \mathbf{0}$ after all. Hence the only matrices that can have 0 as an eigenvalue are singular matrices.

Is the converse true? Suppose that A is singular. Is it the case that we can find a non-zero vector **x** such that $A\mathbf{x} = \mathbf{0}$? (Of course we always have the *trivial* solution $\mathbf{x} = \mathbf{0}$, for any matrix A, but we are after non-trivial ones.)

The answer is 'yes'. Since A is singular, we know that A does not have full rank and so it will be possible to produce an equivalent system of equations with at least one row of zeros, and since we then have fewer equations than unknowns, we will be able to assign an arbitrary value to some unknown, x_i, and solve for the other variables. In particular, A will have eigenvectors with eigenvalue $\lambda = 0$.

We will move on to examples to demonstrate this theory shortly, but this special case is the signpost for solving the general eigenproblem for a square $n \times n$ matrix A, which is the call to find all scalars λ such that for some non-zero vector **x**, we have $A\mathbf{x} = \lambda \mathbf{x}$. By writing $\lambda \mathbf{x}$ as $\lambda I \mathbf{x}$, where I is the $n \times n$ identity matrix, we may write this as a single matrix equation:

$$A\mathbf{x} - \lambda I \mathbf{x} = \mathbf{0} \Leftrightarrow (A - \lambda I)\mathbf{x} = \mathbf{0}. \tag{38}$$

This reduces the general problem to the one we have just solved, which is always a delight to a mathematician's eye, for (38) says that λ is an eigenvalue of A if and only if 0 is an eigenvalue of the matrix $A - \lambda I$, which we know occurs exactly when $A - \lambda I$ is a singular matrix.

Determinants now come into their own, for a square matrix A is singular exactly when its determinant is zero, and so we have the definitive conclusion: the eigenvalues λ of A are the real solutions (if any) to the equation $|A - \lambda I| = 0$. This equation, known as the *characteristic equation of A*, is a polynomial equation of degree n in the unknown λ. Notice that the matrix λI is simply a copy of the identity matrix I with λ replacing 1 all the way down the principal diagonal. Hence the matrix $A - \lambda I$ is obtained by subtracting the symbol λ from all the entries of the principal diagonal of A.

Having developed the basic theory, let's do a typical example: find the eigenvalues and corresponding eigenvectors of the matrix

$$A = \begin{bmatrix} 2 & 7 \\ -1 & -6 \end{bmatrix}.$$

We need to solve the quadratic equation

$$\begin{vmatrix} 2-\lambda & 7 \\ -1 & -6-\lambda \end{vmatrix} = 0 \Leftrightarrow -(2-\lambda)(6+\lambda) - (7)(-1) = 0$$

$$\Rightarrow (\lambda - 2)(\lambda + 6) + 7 = 0 \Rightarrow \lambda^2 - 2\lambda + 6\lambda - 12 + 7 = 0$$

$$\Rightarrow \lambda^2 + 4\lambda - 5 = 0 \Rightarrow (\lambda + 5)(\lambda - 1) = 0;$$

and so our two eigenvalues are $\lambda = -5$ and $\lambda = 1$. To find the eigenvectors that go with the eigenvalues λ, we need to solve the system of equations $A\mathbf{x} = \lambda\mathbf{x} \Leftrightarrow (A - \lambda I)\mathbf{x} = \mathbf{0}$. In this example, for $\lambda = -5$ we get

$$\begin{bmatrix} (2-(-5)) & 7 \\ -1 & -6-(-5) \end{bmatrix} \begin{bmatrix} x \\ y \end{bmatrix} = \begin{bmatrix} 0 \\ 0 \end{bmatrix}$$

$$\Leftrightarrow \begin{bmatrix} 7 & 7 \\ -1 & -1 \end{bmatrix} \begin{bmatrix} x \\ y \end{bmatrix} = \begin{bmatrix} 0 \\ 0 \end{bmatrix}$$

$$\Leftrightarrow \begin{bmatrix} 1 & 1 \\ 1 & 1 \end{bmatrix} \begin{bmatrix} x \\ y \end{bmatrix} = \begin{bmatrix} 0 \\ 0 \end{bmatrix},$$

where we have divided the first row by 7 and the second by -1. The two equations represented by the system have then turned out to be the same, but such redundancy had to appear as the eigenvalues necessarily have the effect of introducing a coefficient matrix that is not of full rank. We see that $\mathbf{x}$ will satisfy this system provided that $x + y = 0$, so that the eigenvectors with eigenvalue $\lambda = -5$ are exactly the non-zero vectors of the form $(a, -a)^T$; in particular, $(1, -1)$ is such an eigenvector. Similarly, for $\lambda = 1$ the system reduces to the single equation $x + 7y = 0$, so the associated eigenvectors are all non-zero multiples of $\mathbf{x} = (7, -1)^T$.

It is the case, of course, that matrices themselves may satisfy a polynomial equation, and on this topic we shall say but one thing. A square matrix has a unique *minimal polynomial*, that is to say a monic polynomial of least degree that has the matrix itself as a (matrix) root and which is a factor of every other polynomial with this property. This polynomial is always a factor of the characteristic polynomial and, put in this fashion, we have the *Cayley-Hamilton Theorem*: *any square matrix satisfies its characteristic polynomial*. To see this in action, take our current example, whose characteristic polynomial is $\lambda^2 + 4\lambda - 5$. To make sense of the constant term we treat 5 as $5I$, and by zero we mean the zero matrix of the appropriate size. A routine calculation will now verify that it is indeed the case that $A^2 + 4A - 5I = 0$.

Similarity and diagonalization

One big topic in matrix theory concerns the *diagonalization* of square matrices, which takes on a particularly simple form in the case of an $n \times n$ matrix A with n distinct eigenvalues. Explaining this topic allows us to touch on another fundamental idea in linear algebra, which is that of similarity of matrices: two matrices A and B are *similar* if there is some invertible matrix P such that $A = PBP^{-1}$. Similar matrices are genuinely similar, in that they represent the same linear transformation of n-space but with respect to different bases. They also share the same determinant, because if $A = P^{-1}BP$ then

$$|A| = |P^{-1}BP| = \frac{1}{|P|} \times |B| \times |P| = |B|.$$

And they have common eigenvalues too, for if λ is an eigenvalue of B with eigenvector $\mathbf{x}$ then $P\mathbf{x}$ is an eigenvector of A with the same eigenvalue:

$$A(P\mathbf{x}) = PBP^{-1}P\mathbf{x} = PBI\mathbf{x} = PB\mathbf{x} = P(\lambda\mathbf{x}) = \lambda(P\mathbf{x});$$

and we may reverse this line of argument to show that if $\mathbf{x}$ is an eigenvector of A then $P^{-1}\mathbf{x}$ is an eigenvector of B with the same eigenvalue.

Suppose now that A is an $n \times n$ matrix with n distinct, real eigenvalues $\lambda_1 < \lambda_1 < \ldots < \lambda_n$, and let D be the diagonal matrix $D = \text{diag}(\lambda_1, \lambda_2, \ldots, \lambda_n)$. The matrices A and D are then similar. The matrix P involved in the similarity of the pair of matrices A and D is the one whose jth column is an eigenvector $\mathbf{x}_j$ for the eigenvalue λ_j, so that it is natural to write $P = [\mathbf{x}_1|\mathbf{x}_2|\ldots|\mathbf{x}_n]$. It now follows from a routine calculation that $AP = PD$ or, what is the same, $A = PDP^{-1}$. (The significance of all the eigenvalues being different from one another is that this ensures that the matrix P is of full rank, and so P^{-1} exists.)

We finish with an application of this result, that of finding powers of a square matrix. We saw in Chapter 7 in our network example that the taking of powers is an operation that often arises in practical calculations. Indeed, in many mathematical models, a process is represented by the action of a matrix A on a vector of variables. Typically, this process could represent a commercial or a biological cycle of some kind. Iteration of the cycle reveals the long-term evolution of the process, and that corresponds to taking powers of A.

If we can diagonalize the matrix, this is a simple thing to calculate because

$$A^k = (PDP^{-1})(PDP^{-1})\ldots(PDP^{-1}),$$

where there are k factors on the right. However, within this product, each P^{-1} is followed by a P, thus giving a factor of $P^{-1}P = I$, the identity matrix, which can be dropped. Therefore all the internal instances of P^{-1} and P vanish and the product telescopes to the simpler product PD^kP^{-1}. The bonus here is that it is trivial to compute powers of a diagonal matrix, as $D^k = \text{diag}(\lambda_1^k, \lambda_2^k, \ldots, \lambda_n^k)$. In effect, the k-fold matrix product has been converted into a product of just three matrices.

Negative powers as well are available:

$$A^{-k} = (A^k)^{-1} = (PD^kP^{-1})^{-1} = (P^{-1})^{-1}D^{-k}P^{-1} = PD^{-k}P^{-1}.$$

The determinant of a diagonal matrix D is just the product of the entries of its principal diagonal. (In fact, this is true of any *lower* or *upper triangular* matrix: a lower triangular matrix has only zero entries above its leading diagonal, and an upper triangular matrix has only zeros below). Since similar matrices share the same determinant, we may conclude therefore that the determinant of an $n \times n$ matrix with n distinct eigenvalues equals the product of those eigenvalues (a result that holds generally if we allow for complex and repeated roots of the characteristic equation).

An open-ended topic in matrix theory concerns the factorization of matrices. A visit to the Internet will provide a host of examples. The *LU (lower-upper) factorization* applies to square matrices: $A = LU$, where L is lower triangular while U is upper triangular. Alternatively, there is also the *QR factorization* (or *decomposition*) $A = QR$, where R is again upper triangular and Q is *orthogonal*, meaning Q is a square matrix whose inverse equals its transpose, so that $QQ^T = I$. The *QR Algorithm* for the computation of eigenvalues, which is based on the QR decomposition, is one of the most important algorithms in mathematics and was discovered independently by John G. F. Francis in England and the Soviet mathematician Vera Kublanovskaya in 1961. Although with matrix factorization we do not have a theorem that corresponds directly to the prime factorization of integers, the fact that the action of a matrix may be split into the action of two or more matrices with special properties is useful in a wide variety of applications.

Chapter 10
Vector spaces

More on abelian groups

This chapter features the algebra of vector spaces, which are abelian groups with an additional scalar multiplication by a field. By way of preamble, we look again at certain aspects of abelian groups.

One example type that we met in Chapter 6 was that of the cyclic group, $(\mathbb{Z}_n, +)$. We can generate another abelian group by placing two or more cyclic groups in parallel in the following fashion. Consider all pairs of the form (i, j) where i and j are drawn from the cyclic groups $\mathbb{Z}_m$ and $\mathbb{Z}_n$, respectively: this set is written as $G = \mathbb{Z}_m \times \mathbb{Z}_n$, and we go on to define an addition on this *direct product* by adding components separately, with the first entry operating modulo m and the second modulo n. This gives a new abelian group with $m \times n$ members in all.

The simplest example is when $m = n = 2$, for then G has $2 \times 2 = 4$ members, and $G = \{(0, 0), (1, 0), (0, 1), (1, 1)\}$. An example of an addition is $(0, 1) + (1, 1) = (1, 0)$, as for the second entry we get $1 + 1 \equiv 0 \pmod{2}$. It is simple to check that $(0, 0)$ is the identity element and that for this particular group G, each member is self-inverse. Furthermore, the sum of any two distinct non-zero members of G is equal to the third.

It turns out that these are the only finite commutative groups: every finite abelian group is a direct product of cyclic groups, although in general we have to allow for more than two cyclic groups in the product. There is a pair of structure theorems for finite abelian groups. Any finite abelian group can be represented in one of two special ways based on numerical relationships between the subscripts of the cyclic groups involved. In one representation, we make all the subscripts powers of primes, in the alternative, we make each subscript a divisor of its successor. In this way we may decide when two finite abelian groups are isomorphic, which is to say essentially the same, by checking whether or not their representations are identical when they are presented in either of these two ways.

For example, $G = \mathbb{Z}_2 \times \mathbb{Z}_2$ and $H = \mathbb{Z}_4$ each have four members, but they are not two copies of the same abelian group for, as we have already seen, each member of G is its own inverse but that is not the case for the cyclic group H, as, for instance, $1 + 1 = 2 \not\equiv 0$ (mod 4). In contrast, $G = \mathbb{Z}_2 \times \mathbb{Z}_3$ is isomorphic to $\mathbb{Z}_6$ as G is cyclically generated by the pair $(1, 1)$: by successively adding $(1, 1)$ to itself, we obtain in turn all six members of G, namely $(1, 1), (0, 2), (1, 0), (0, 1), (1, 2), (0, 0)$, and so G and H both represent a 6-cycle.

In general, $\mathbb{Z}_m \times \mathbb{Z}_n$ is isomorphic to $\mathbb{Z}_{mn}$ if m and n are relatively prime, but not otherwise. This is a modern interpretation of what is called the *Chinese Remainder Theorem*, which says that the system of two (or more) simultaneous congruences in relatively prime moduli of the form $x \equiv a \pmod{m}$ and $x \equiv b \pmod{n}$ has a unique solution modulo mn.

In the classical texts, a typical problem might ask for the smallest number that leaves a remainder of 1 when divided by 4 and 8 when divided by 9. This in effect asks for the member n of $\mathbb{Z}_{36}$ that corresponds to $(1, 8)$ in $\mathbb{Z}_4 \times \mathbb{Z}_9$. Since $n \equiv 1 \pmod{4}$, we may put $n = 1 + 4t$. We also require that $n \equiv 8 \pmod{9}$, and so we substitute accordingly into this second congruence to obtain

$$1 + 4t \equiv 8 \pmod{9}$$

$$\Rightarrow 4t \equiv 7 \equiv 16 \pmod{9}$$

$$\Rightarrow t \equiv 4 \pmod{9},$$

and so we arrive at $n = 1 + 4t = 1 + 4 \times 4 = 17$. This is indeed the solution as in $\mathbb{Z}_4 \times \mathbb{Z}_9$ we see that $17(1, 1) = (1, 8)$.

Vector spaces

Vector spaces are often the first type of axiomatically defined algebra with which university students are presented. The notion was formally defined in 1888 by Giuseppe Peano (1858–1932) of Turin and became central to mainstream mathematics from around 1920. Peano himself was led by the earlier efforts of the German scholar Hermann Grassmann (1809–77).

Vector spaces form the backdrop of linear mathematics, which underpins not only applied algebra but also aspects of the mathematics of the continuous. In short, once on the lookout for vector spaces you discover that they are common throughout mathematics. They are also a good place to learn how to frame abstract algebraic arguments, partly because there are natural examples to call upon that are already within a student's experience, but also because many of the important proofs are quite simple.

A *vector space* $(V, +)$ is first and foremost an abelian group. Often, but not always, V is an infinite group and the prototypes are the abelian groups $\mathbb{R}^2 = \mathbb{R} \times \mathbb{R}$, $\mathbb{R}^3$, and more generally $\mathbb{R}^n$, which consists of n-tuples $(a_1, a_2, \ldots, a_n)$ of real numbers, added by components (vector addition, which is just a special case of matrix addition). These vectors can themselves be multiplied by scalars, which are typically members of the field $\mathbb{R}$. This is the second aspect of vector spaces: for each there is an associated *field of scalars*, F, and vectors $\mathbf{u}$, $\mathbf{v}$ may be multiplied by scalars a, b to produce another vector, subject to the following laws, which are kinds of mixed associativity and distributivity laws, all of which we have seen in (28) when considering matrix multiplication by scalars:

$$a(b\mathbf{v}) = (ab)\mathbf{v}, \quad a(\mathbf{u}+\mathbf{v}) = a\mathbf{u} + b\mathbf{v}, \quad (a+b)\mathbf{u} = a\mathbf{u} + b\mathbf{u}.$$

In addition, we need to insist that $1\mathbf{v} = \mathbf{v}$. This is not a consequence of the other laws: without it we could for example have $a\mathbf{v} = \mathbf{0}$ for every scalar a and vector $\mathbf{v}$, and our scalar multiplication laws would all hold, albeit in a trivial way.

Given these rules we may, in a similar fashion to the rule about zero multiplication in Chapter 2, deduce that $0\mathbf{v} = \mathbf{0}$ (the zero on the left is the zero of the scalar field and that on the right the $\mathbf{0}$ of V). From this we may deduce further that $-1\mathbf{v} = -\mathbf{v}$ as follows:

$$\mathbf{0} = 0\mathbf{v} = (1+(-1))\mathbf{v} = 1\mathbf{v} + (-1)\mathbf{v} = \mathbf{v} + (-1)\mathbf{v}$$
$$\Rightarrow (-1)\mathbf{v} = -\mathbf{v},$$

where the final conclusion invokes the uniqueness of inverses in a group. Note also that along the way we did indeed call upon the axiom that $1\mathbf{v} = \mathbf{v}$.

Certainly $\mathbb{R}^n$ satisfies the requirements of a vector space, but to find another less obvious example we look to the solution set S of a set of simultaneous equations, which we write in matrix form as $A\mathbf{x} = \mathbf{0}$. Let us assume that $\mathbf{x}$ consists of n unknowns so that S is a subset of $\mathbb{R}^n$, and since we have a *homogeneous system*, meaning that the RHS is zero, we know that S contains at least the solution $\mathbf{x} = \mathbf{0}$. However, S may contain more solutions: if we look at the 2×3 system (9) that arose in the Lincoln Fair problem of Chapter 3, we find that there is a single infinity of solutions to that system in that one variable can be freely assigned any value and the remaining two variables may then be expressed in terms of the free variable.

Moreover, the solution set S is a vector space in its own right. There is certainly no trouble about S satisfying the equational vector space axioms, as these properties are all known to hold in the larger vector space $\mathbb{R}^n$. What needs to be looked at, however, is whether or not S is closed under the vector space operations of addition and scalar multiplication (and that S is not empty, which we have already checked).

We can do this in one action by verifying an equivalent criterion, that being that S is closed under the taking of *linear combinations*, meaning that for all scalars, a and b, and all members **u**, **v** of S, $a\mathbf{u} + b\mathbf{v}$ is also in S. But this follows at once by standard properties of matrices:

$$A(a\mathbf{u} + b\mathbf{v}) = aA\mathbf{u} + bA\mathbf{v} = a\mathbf{0} + b\mathbf{0} = \mathbf{0} + \mathbf{0} = \mathbf{0},$$

and so the solution set S of a homogeneous system of equations in $\mathbb{R}^n$ is a *subspace* of $\mathbb{R}^n$.

This has consequences for the more general system $A\mathbf{x} = \mathbf{b}$, for if **a** is *any* particular solution of this latter system, so that $A\mathbf{a} = \mathbf{b}$, then the totality of all solutions is given by $S + \mathbf{a}$, meaning that if **x** lies in S then $\mathbf{x} + \mathbf{a}$ is a solution to $A\mathbf{x} = \mathbf{b}$ (which follows at once, as then $A(\mathbf{x} + \mathbf{a}) = A\mathbf{x} + A\mathbf{a} = \mathbf{0} + \mathbf{b} = \mathbf{b}$), and that every solution **c** to the general system has this form. To see, just note that $\mathbf{x} = \mathbf{c} - \mathbf{a}$ solves $A\mathbf{x} = \mathbf{0}$, as

$$A\mathbf{x} = A(\mathbf{c} - \mathbf{a}) = A\mathbf{c} - A\mathbf{a} = \mathbf{b} - \mathbf{b} = \mathbf{0};$$

hence **x** is in S, and so the typical solution $\mathbf{c} = \mathbf{x} + \mathbf{a}$ does indeed lie in $S + \mathbf{a}$.

This approach of expressing the general solution to a system of equations as a translate of the solutions of a homogeneous system that forms a subspace of a related vector space is also applicable to quite different kinds of equations, including the systems of so-called linear differential equations that typically arise in mathematical models of moving physical systems.

As has already been touched upon, the 'dimension' of the solution set for our pair of simultaneous equations from our Lincoln Fair problem is 1, in that there is one degree of freedom on offer when specifying a particular trio of numbers that satisfy both equations. This leads us to the question as to what we mean by the dimension of a vector space.

If we take any subset A of a vector space V, then the collection U of all possible linear combinations of members of A is itself a subspace of V, as U is closed under the taking of linear combinations. Since any subspace that contains A necessarily contains all such linear combinations, it follows that U is the smallest subspace of V that contains A: we say that U is the *subspace generated by A* and that A is a *spanning set* for the subspace U.

Next, a subset $I = \{\mathbf{u}_1, \mathbf{u}_2, \ldots, \mathbf{u}_k\}$ of V is called *independent* if no member of I is a linear combination of other members of I. An equivalent definition is that

$$(a_1\mathbf{u}_1 + a_2\mathbf{u}_2 + \ldots + a_k\mathbf{u}_k = \mathbf{0}) \Rightarrow (a_1 = a_2 = \ldots = a_k = 0). \quad (39)$$

This does say the same thing—if the condition of this definition is violated, then some $a_i \neq 0$ and we may then make $\mathbf{u}_i$ the subject of the equation on the left, while conversely, if $\mathbf{u}_i$ is a linear combination of other vectors from I, then taking all terms to the LHS gives a linear combination that is equal to $\mathbf{0}$ yet the coefficient of $\mathbf{u}_i$ equal to 1 (and so $a_i = 1 \neq 0$).

The definition (39) may seem a more technical formulation of the idea of independence but, because it is framed in terms of equations, it lends itself to algebraic manipulation and so is useful in mathematical argument. Furthermore, the condition that *some*

$\mathbf{u}_i$ is a linear combination of the other vectors does *not* imply that *every* vector in the list is a linear combination of the others: for a simple example in $\mathbb{R}^2$ that makes this point, we may take $\mathbf{u}_1 = (0, 0)$ and $\mathbf{u}_2 = (1, 0)$. Then $\mathbf{u}_1 = 0\mathbf{u}_2$, so that the set $\{\mathbf{u}_1, \mathbf{u}_2\}$ is not independent, but $\mathbf{u}_2$ is not a multiple of $\mathbf{u}_1$.

The importance of independent sets lies in the fact that they can be used to coordinatize vector spaces, for suppose that I is independent and U is the subspace of V generated by I. Then I is both independent and a spanning set for U, and as such is known as a *basis* for the vector space U. Any member $\mathbf{u}$ of U can of course be written as a linear combination of the members of I because I spans U but, crucially, that linear combination is unique, for suppose we had two such representations:

$$\mathbf{u} = a_1\mathbf{u}_1 + a_2\mathbf{u}_2 + \ldots + a_k\mathbf{u}_k = b_1\mathbf{u}_1 + b_2\mathbf{u}_2 + \ldots + b_k\mathbf{u}_k.$$

Then, by subtraction, we obtain

$$(a_1 - b_1)\mathbf{u}_1 + (a_2 - b_2)\mathbf{u}_2 + \ldots + (a_k - b_k)\mathbf{u}_k = \mathbf{0},$$

and since I is independent, this implies from our definition (39) that *all* these coefficients equal zero, which is just to say that $a_1 = b_1, a_2 = b_2, \ldots, a_k = b_k$. Therefore the representation of $\mathbf{u}$ in terms of the members of I is unique.

This coordinatization allows us to identify $\mathbf{u}$ with the corresponding list of coordinates $(a_1, a_2, \ldots, a_k)$, and the associated mapping is an isomorphism of U onto the vector space F^k, where F is the scalar field of multipliers of our vector space V. That is to say, regarded as a vector space, U is just a copy of F^k.

A vector space generally possesses lots of bases. When dealing with $\mathbb{R}^3$, for instance, we have favoured the *standard basis*, $\mathbf{b}_1 = (1, 0, 0)$, $\mathbf{b}_2 = (0, 1, 0)$, $\mathbf{b}_3 = (0, 0, 1)$. This trio of vectors has the additional nice properties of all being of length 1, and each being perpendicular to the other two. However, *any* three vectors in $\mathbb{R}^3$ form a basis for $\mathbb{R}^3$ as long as none is a linear combination of the others, which amounts to saying that the corresponding position vectors do not all lie in the one plane. Particular problems sometimes may be simplified using bases other than the standard basis, such as those involving eigenvectors of the associated matrix.

It is the case, however, that every vector space possesses a basis and that any two bases have the same size. This common size is called the *dimension* of the vector space. Some vector spaces are, however, infinite-dimensional. For example, the collection of all polynomials with real coefficients form a vector space under polynomial addition, and a natural basis here is the list of all powers of x, i.e. $1, x, x^2, \ldots, x^n, \ldots$, which is clearly infinite in extent.

The proof that all bases of a vector space V have a common size depends on what is known as the *Exchange Lemma*, which has as a consequence that any independent set I of V is never larger in size than any spanning set S of V. If we then have two bases of V, B_1 and B_2, then B_1 is no larger than B_2, as B_1 is independent and B_2 spans V, and, by the same reasoning, B_2 is no larger than B_1. Therefore the two bases must have the same size, the dimension of V.

Another significant consequence of the Exchange Lemma is that every independent set I of V may be extended to a basis B of V, meaning that B contains I as a subset. In particular, a basis I of a subspace U of a vector space V can be extended to a full basis B of V. It follows from this that the dimension of a subpace is always less than or equal to that of the containing vector space V, and indeed we can only have equality in their dimensions if $U = V$.

Many important facts about matrices can be proved by applying these ideas to the *row space* and the *column space* of an $m \times n$ matrix A, which are the respective subspaces of $\mathbb{R}^n$ and $\mathbb{R}^m$

spanned by the rows of A and the columns of A. In particular, it can be proved that the row rank and column rank have a common value, which is therefore called the *rank* of A, and that for any matrix product AB, $\text{rank}(AB) \leq \text{rank}(A)$.

Finite fields

In this final section, we bring together the ideas of modular arithmetic, the construction of the complex numbers, factorization of polynomials, and vector spaces to explain the existence of finite fields, closing the discussion with a concrete example. We have already noted that the commutative ring $(\mathbb{Z}_p, +, \cdot)$, where p is prime, is indeed a field of p elements. There are finite fields, however, that are not of this simple type. Nonetheless, the field laws coupled with finiteness impose a great deal of restriction on the structure of any finite field, which is why there are so few of them. Indeed, for every prime number p and positive integer n, there is, up to isomorphism, just one finite field with p^n members, and there are no others.

The highly structured nature of finite fields has allowed many applications not only to problems in algebra but also to modern cryptography, where the difficulty of the so-called discrete logarithm problem in finite fields or in elliptic curves is the basis of common protocols, such as the Diffie–Hellman protocol. In coding theory, many codes are constructed as subspaces of vector spaces over finite fields. In pure number theory, the famous proof by Sir Andrew Wiles of Fermat's Last Theorem involved a host of sophisticated mathematical tools, finite field theory being among them. Let us look at how this delicate structure comes about.

A finite field, is firstly, a finite abelian group $(F, +)$, and from this it can be deduced that there is a least positive integer, p, such that for every member a of F we get $pa = a + a + \cdots + a = 0$ (where there are p instances of a in the sum). What is more, it can be shown that p is itself a prime and, indeed, F has a copy of the field

($\mathbb{Z}_p$, +, ·) embedded within it. If we use the number 1 to denote the multiplicative identity of F, then this *prime subfield* of F, can be written as $\mathbb{Z}_p = \{0, 1, 2, \ldots, p-1\}$.

We may now note further that F is a vector space over its prime subfield (the axioms hold through F being a field) and so F has a basis B, of let us say n elements. Each member of F can then be written uniquely as a linear combination of the n members of B, giving a total of p^n elements of F. To reiterate, any finite field has p^n elements for some prime p and some positive integer n.

Both $(F, +)$ and $(F \setminus \{0\}, \cdot)$ are finite abelian groups. The additive group of F is simply the direct product of n copies of $\mathbb{Z}_p$. The multiplicative group is even simpler, being the cyclic group $\mathbb{Z}_q$, where $q = p^n - 1$ (of course, 0 is not a member of this group). This can be proved through use of the structure theorems for finite abelian groups mentioned in the first section of this chapter, and by analysis of the roots of a polynomial equation of the form $x^m = 1$, as there is a common power m of all the non-zero members of F that gives 1. The finite field with p^n elements is the smallest field in which the polynomial $x^{p^n} = x$ has p^n distinct roots.

As an example, we construct the finite field F_9, with $3^2 = 9$ elements. By the comments above, the prime subfield of F_9 will be $\mathbb{Z}_3 = \{0, 1, 2\}$ under addition and multiplication modulo 3. In this field, we have of course that $2 = -1$, for $1 + 2 = 3 \equiv 0 \pmod{3}$. We now note that $\mathbb{Z}_3$ lacks a square root of -1, for $0^2 = 0$, $1^2 = 1$, and $2^2 = 4 \equiv 1 \pmod{3}$. In a manner reminiscent of the way in which we constructed the field of complex numbers, we remedy this situation by introducing a new symbol, a, endowed with the property that $a^2 = -1$, and adopt a as a member of F_9. This leads to the analogue of the complex numbers over the field $\mathbb{Z}_3$, giving us $3^2 = 9$ 'complex' numbers, those being the nine formal sums $a + ba$, where a and b are members of $\mathbb{Z}_3$. We now combine them according to the field laws, making use of the equation $a^2 = -1 = 2$ whenever powers of a arise.

Table 2. Product table for the eight non-zero members of the nine-element field.

·	**1**	**2**	**α**	**1 + α**	**2 + α**	**2α**	**1 + 2α**	**2 + 2α**
1	1	2	a	$1 + a$	$2 + a$	$2a$	$1 + 2a$	$2 + 2a$
2	2	1	$2a$	$2 + 2a$	$1 + 2a$	a	$2 + a$	$1 + a$
α	a	$2a$	2	$2 + a$	$2 + 2a$	1	$1 + a$	$1 + 2a$
1 + α	$1 + a$	$2 + 2a$	$2 + a$	$2a$	1	$1 + 2a$	2	a
2 + α	$2 + a$	$1 + 2a$	$2 + 2a$	1	a	$1 + a$	$2a$	2
2α	$2a$	a	1	$1 + 2a$	$1 + a$	2	$2 + 2a$	$2 + a$
1 + 2α	$1 + 2a$	$2 + a$	$1 + a$	2	$2a$	$2 + 2a$	a	1
2 + 2α	$2 + 2a$	$1 + a$	$1 + 2a$	a	2	$2 + a$	1	$2a$

The addition table of F_9 is that of $\mathbb{Z}_3 \times \mathbb{Z}_3$: for example,

$$(1 + a) + (2 + a) = 3 + 2a = 0 - a = -a.$$

The multiplication table is more interesting, but is now easily compiled, if we remember to replace each instance of a^2 by -1 (or 2) as we proceed. The multiplicative group of non-zero members of F_9 is as shown in Table 2.

Sample calculations:

$$2a(2 + a) = 4a + 2a^2 = a + 2(-1) = a - 2 = 1 + a;$$
$$(2 + a)^2 = 4 + 4a + a^2 = (4 - 1) + a = 3 + a = a.$$

Being a group table, Table 2 has the Latin square property, in that each symbol of the group appears exactly once in each row and each column of the body of the table. Being an abelian group table, it is symmetric with respect to reflection in the principal diagonal of the table. In accord with our description above, this group is the cyclic group $\mathbb{Z}_8$, having, for example, $1 + a$ as a generator, meaning that the eight powers of $1 + a$ comprise the entire group:

$$(1+a), \quad (1+a)^2 = 2a, \quad (1+a)^3 = 1+2a, \quad (1+a)^4 = 2,$$
$$(1+a)^5 = 2+2a, \quad (1+a)^6 = a, \quad (1+a)^7 = 2+a, \quad (1+a)^8 = 1.$$

And with this, one of the prettiest of multiplication tables, we end our book.

Further reading

This book provides no more than a rapid and skeletal introduction to modern algebra. To put further flesh on the bones requires more familiarity with other aspects of mathematics, particularly calculus, for the development of algebra has not taken place in a vacuum. For instance, matrix algebra is central to the calculus of several variables, and familiarity with one develops comprehension of the other. Indeed, whole areas of mathematics, such as topology, have an algebraic side to them.

For a careful and thorough introduction to algebra, the reader may go to any textbook with a title along the lines of 'Introduction to Abstract/Modern Algebra'. Books on discrete mathematics will also involve algebra in a substantial way, while focusing on applications to networks. Before that, however, the reader may be advised to learn more about *linear algebra* as, along with calculus, the algebra of matrices lies at the core of advanced mathematics. Moreover, an introduction to vector spaces, which arises through abstraction from linear algebra, is perhaps the gentlest introduction to abstract algebra. Group theory does indeed lie at the heart of contemporary algebra, but the theory of groups can be a difficult topic unless the student has some prior experience of working with more concrete algebraic types.

An online search for algebra books will immediately yield plenty of examples, and it is not hard to tell the difference between those that focus on school algebra of the kind read in the first four chapters of this

VSI and those designed for university work. For those reasons, I will not make a list here of the many excellent books on modern algebra.

However, if the reader would like to learn more on the human side of the subject, there is for example *Taming the Unknown: A History of Algebra from Antiquity to the Early Twentieth Century* by Victor J. Katz and Karen Hunger Parshall (Princeton: Princeton University Press 2014). A nice historical introduction to complex numbers is *An Imaginary Tale: The Story of $\sqrt{-1}$* by Paul J. Nahin (Princeton: Princeton University Press 2010). I also make special mention of the book *Abel's Proof: An Essay on the Sources and Meaning of Mathematical Unsolvability* by Peter Pesic (Cambridge MA: MIT Press 2004). This is not an algebra text but it does take the reader through the history of the unsolvability of the quintic and, in detail, explains the original proof of Ruffini and Abel as opposed to that provided through Galois theory, which is the basis of the modern theory of polynomial equations.

Index

A

absolute value 35–6
addition
  of complex numbers 58
  modulo n 82
additive
  identity 14, 23
  inverse 14
algebraic identity 16
Argand plane 58
axis
  imaginary 59
  real 59

B

binary operation 75
Binomial
  coefficients 19
  Theorem 18–20, 78

C

Cayley-Hamilton Theorem 122
Celsius 28–9
Chinese Remainder Theorem 127
coefficient 25, 53
  leading 53
column space 133
coordinate plane 27
completing the square 40, 42–3
complex number *see also* number
  conjugate of 61
  modulus of 61
  polar form of 63
composition (of mappings) 97
congruence modulo n 80
  linear 84–6
conjugate 60
Conjugate Root Theorem 68
constant term 53
constraint 32
Cramer's Rule 118
cross-term 16, 43

D

decimal 57
  recurring 57
denominator 15
  common 15
  rationalizing 60
difference of two squares 17
  of two cubes 17
discriminant of
  cubic 72
  quadratic 50–2
Division Algorithm 8

E

eigenvalue 118–22
eigenvector 118–22
elimination 31–2
    Jordan-Gaussian 118
empty set 21
equation(s) 16
    characteristic 121
    cubic 69
    depressed cubic 70
    homogeneous system of 129
    linear 25, 27
    monic 43
    quadratic 40
    quartic 73
    quintic 74
    simultaneous 25, 30–1, 37–8, 89
    slope-intercept form 30
    substitution 30–2, 39
equivalence class 80
Euclidean
    Algorithm 8
    Lemma 7, 65
Exchange Lemma 133

F

factor 7, 56
Factor Theorem 56–7, 63
factorials 20
factorization 6
Fahrenheit 28–9
field 76
    finite 134–41
    of fractions 79
    prime subfield 135
    of scalars 129
function 54
    linear 27
Fundamental Theorem of Algebra 67–8
Fundamental Theorem of Arithmetic 7

G

Golden Rectangle 48–9
Golden Ratio 49, 66
gradient 27
greatest common divisor 8
group 75, 103
    abelian 76, 126–9
    cyclic 103, 126
    dihedral 104
    direct product of 126
    isomorphism 105, 127
    Latin square property 105, 136
    presentation by generators and relations 104
    subgroup of a 103
    of symmetries 104

H

highest common factor 8

I

identity element (of a group) 75
implies sign 23
independent set 91
indices 18
inequalities 25, 32–4
    signs 6, 32
integral domain 78
intercepts 27
inverse 14, 75

L

lattice 76
laws
    associative 12, 75
    commutative 13, 76
    distributive 13
    of indices 18
least residues 80
line
    equation of 27, 30

linear
  algebra 94
  combination 88, 91, 130
  mapping 96
linear programming 39
logarithm 19

M

matrix 87
  addition 88
  augmented 108
  characteristic equation of 121
  coefficient 89
  co-factor of 115
  column 87
  column space of 133
  determinant of 89, 99, 109, 111–17, 121
  diagonal 94
  diagonalization of 123–5
  eigenvalue of 118–22
  eigenvector of 118–22
  elementary row matrix 107
  identity 101
  incidence 93
  integer 109, 116, 117–18
  inverse 102
  invertible 109
  Kirchhoff 94, 115
  lower triangular 124–5
  LU factorization of 125
  minimal polynomial of 122
  minor 113
  multiplication 90
  normalizing condition 117
  orthogonal 125
  pivot(ing) 108
  principal diagonal of 94
  QR factorization of 125
  rank (row or column) 91
  row 87
  row space of 133
  similar 123–5
  singular 102, 120
  square 87
  symmetric 93
  transpose of 106
  upper triangular 124–5
  zero 102
mean
  arithmetic 36
  geometric 36
modular arithmetic 79–84
module 76
multiplication
  of complex numbers 60
  of matrices 90
  modulo n 82
multiplicative
  identity 14, 23
  inverse 22

N

node
  degree of 94
network 92
number(s)
  complex 57–61
  composite 7
  divisor of 7
  factorization of 7
  imaginary 57
  integer 4
  irrational 57, 65–6
  line 4
  multiple of 7
  natural 3
  negative 3
  prime 7
  prime factorization of 7
  positive 4
  rational 4, 57
  real 57
  relatively prime 8, 127
numerator 15

P

parabola 44–5, 50

parallelepiped 112
permutation 114–15
plane
  Argand 58
  equation of 37
polynomial 53
  cubic 53
  degree of 53
  quotient 55
  roots of 41, 49–50, 56
powers 6, 18
  of a matrix 124
prime factorization 7
product
  dot 90
  scalar 90, 91

Q

quadratic *see also* equation
  formula 40, 46–8
  irreducible 69

R

Rational Root Theorem 65
rationalizing the denominator 60
remainder 7, 54, 80
ring 76
  commutative 77
  of integers 77
  of integers mod n 79
  unital 77
roots *see also* polynomial
  multiplicity of 69
row operations on matrices 107
row space 133

S

scalar multiplication 88
semigroup 76
slope 27

square root 2, 18, 34–5
subset 4
subtraction 3, 5
surd 60
summation notation 20

T

temperature
  Celsius 28
  Farenheit 28
transformation 96
  generated by 101
  identity 102
  linear 96
transitive property 80
tree 95
  spanning 95, 115–16

U

universal algebra 76

V

vector(s)
  addition 59
  basis 97
  column 90
  eigenvector 118
  independent set of 131
  position 95
  row 90
  zero 91
vector spaces 128–41
  basis of 132
  dimension of 133
  independent subset of 131
  spanning set of 131
  standard basis of 132
  subspace generated by 131
  subspace of 130
Vieta substitution 70–2